NOBEL LAUREATES
IN SEARCH OF

IDENTITY &

INTEGRITY
Voices *of* Different Cultures

NOBEL LAUREATES
IN SEARCH OF

IDENTITY &

INTEGRITY
Voices *of* Different Cultures

Editor

Anders Hallengren
Stockholm University, Sweden

World Scientific

NEW JERSEY · LONDON · SINGAPORE · BEIJING · SHANGHAI · HONG KONG · TAIPEI · CHENNAI

Published by

World Scientific Publishing Co. Pte. Ltd.

5 Toh Tuck Link, Singapore 596224

USA office: 27 Warren Street, Suite 401-402, Hackensack, NJ 07601

UK office: 57 Shelton Street, Covent Garden, London WC2H 9HE

Library of Congress Cataloging-in-Publication Data
Hallengren, Anders, 1950–
   Nobel laureates in search of identity and integrity: voices of different cultures /
    edited by Anders Hallengren.
     p. cm.
   A collection of essays, biographies, and Nobel lectures, by and about ten Nobel laureates:
   V.S. Naipaul, Nadine Gordimer, Derek Walcott, Naguib Mahfouz, Patrick White,
   Ernest Hemingway, Grazia Deledda, Amartya Sen, Rabindranath Tagore, Nelson Mandela.
   Includes bibliographical references.
   ISBN-13 978-981-256-038-4 -- ISBN-10 981-256-038-6
   ISBN-13 978-981-256-074-2 (pbk) -- ISBN-10 981-256-074-2 (pbk)
   1. Nobel Prizes. 2. Nobel Prizes--Biography. 3. Multiculturalism. 4. Ethnicity.
   5. Cultural property--Protection. I. Title

AS911.N9 N5854 2004
001.4'4'0922--dc22   2004058654

Printed in Singapore

Contents

❋

Introduction

❋

Addressing current questions about unity in diversity in a multicultural world of change, several Nobel Laureates of Literature have also focused on the difference between their identity as an author and their identity as a social being. This aspect of their creative selves makes them living in, as it were, two different and quite distinct worlds. There is a distance between them that is bridged only in the literary work, where these two worlds are intimately connected and presuppose one another. V. S. NAIPAUL, living in England and born in Trinidad, discusses this topic in his Nobel Lecture of 2001, "Two Worlds". His literary domain, encompassing the West Indies as well as India, Africa, America, and the Islamic countries of the East, reflects his own multicultural background, and in some of his most famous texts, the difficulty of settling down or finding a way "home" is the theme. An example is his novel *The Enigma of Arrival*.

In a similar note, NADINE GORDIMER discussed the complex role of the writer in her Nobel Lecture ten years earlier, "Writing and Being". A white woman of Jewish descent, living all her life in South Africa, she early sided with the black liberation movement. Her work features a transcendence of human identity reaching far beyond race and gender. In her novels she identifies with her protagonists, be they white or black, men or women, criminals or saints, making them deeply understandable.

"A Single, Homeless, Circling Satellite" is a line drawn from a poem by DEREK WALCOTT, referring to himself and many other people in the heterogeneous and colonial world of the West Indies. The small Caribbean Island republic of St. Lucia, which gained independence in 1979, had been the focus of a prolonged struggle between the colonial powers, changing hands fourteen times. Europeans, who brought slaves from Africa, Walcott stemming from both, populated it. In this mid-world between the continents, a point of connection and passage between North and South America, Africa and Europe, Derek Walcott's intense search for an identity — by connecting to local as well as foreign traditions — represents an existential struggle of universal significance.

In the novels and short stories of NAGUIB MAHFOUZ, there is a constant seeking for Egyptian identity behind the weft of illusion and reality. Calling himself "a son of two civilizations" — the Egyptian and the Arabic-Islamic — he started his career as a writer by exploring ancient Egyptian history. He did not only do so to understand the contemporary scene or to criticize it in a covert fashion. His aim was to seek the identity of his own country in the space-time of his existence and the sphere of his Self. He also obviously sought for a reliable anchorage in the distant past during years of war, upheaval, and calamity. As an Arabic author, he transcends the limits of Arabic and Moslem tradition, to which he belongs, tracing his heritage and seeking his identity as an Egyptian.

Born in London by Australian parents, the Australian writer PATRICK WHITE is yet another instance of mixed identity, the difficulty of settling down and a feeling of being lost in this world. His autobiographic book *Flaws in the Glass,* describes how he became a skeptic and an endless seeker. The article found in the present collection shows that his main identity as a writer is that of an existential explorer aiming at a deep sense of humanity.

After that, attention is paid to ERNEST HEMINGWAY (USA) in an inquiry into the person behind the popular image and the macho myth; a penetration based on new research and posthumously published writings. Modern scholarship has added immensely to the depth of our understanding of Hemingway, and many new aspects of Hemingway's life and works that were previously obscured by his public image have now emerged into the light. Later biographic research has revealed, behind the macho façade of boxing, bullfighting, big-game hunting and deep-sea fishing he built up, a sensitive and vulnerable mind that was full of contradictions. In Hemingway, sentimentality, sympathy, and empathy are turned inwards, not restrained, but vibrant below and beyond the level of fact and fable. In the new light, he emerges as an identity-seeking and insecure man, unknown to the public.

GRAZIA DELEDDA — the first Italian woman to receive the Nobel Prize for literature — remained the spokeswoman and storyteller of her Sardinian culture even after her removal to Rome. She wrote in Italian, but her mother's tongue was *logudorese sardo,* a local dialect that can be regarded as another Roman language altogether. Her official education lasted only four years and was on the level of primary school, yet her career as an Italian writer started in her teens, when she began writing short stories for a Roman magazine. Deledda found more infamy than fame in her Sardinian village. Suspicion and rumors followed her. Her mother was attacked for being an irresponsible parent; village women burned a magazine and shouted their reproaches. To deflect the shock and anger engendered by her fiction, Deledda published under pseudonyms for a while, but her literary output grew to enormous proportions. For life sticking to her Sardinian identity and origin, her life project was to create a genuine Sardinian literature on her own and become the Italian voice of her island. However, having attained her goal to immortalize the society she came from, Grazia Deledda ended up as one of its severest critics.

AMARTYA SEN, who received The Bank of Sweden Prize in Economic Sciences in Memory of Alfred Nobel 1998, lives in the USA and Great Britain but grew up in Dhaka, the capital of Bangladesh, and spent three of his childhood years in Mandalay, Burma. He was, however, born in Santiniketan, on the campus of Nobel Laureate RABINDRANATH TAGORE's Visva-Bharati (both a school and a college), where his maternal grandfather Kshiti Mohan Sen used to teach Sanskrit as well as ancient and medieval Indian culture. During his education in Dhaka, Amartya Sen was struck by Rabindranath Tagore's approach to cultural diversity in the world: "Whatever we understand and enjoy in human products instantly becomes ours, wherever they might have their origin." Tagore argued against all separatist views, "against the intense consciousness of the separateness of one's own people from others."

This idea is in a particular sense embraced by Peace Prize Laureate NELSON MANDELA, whose fight for the liberation of the black nations of the Republic of South Africa originally was inspired by Indians, with whom he was soon to cooperate. The multiracial vision of Mohandas Karamchand Gandhi played an important role, and the Mahatma had in fact lived and worked in South Africa for many years. By Mandela referred to as "a South African", Gandhi's dream was a lasting influence — "that all the different races commingle and produce a civilization that perhaps the world has not seen."

In this spirit, Nelson Mandela's fight for freedom was guided by an idea of integrated integrity: all nations and all religious groups of the country should have equal rights and preserve their native tongues and cultural characteristics. Thus, when the first "black" government was formed in 1994, it was in reality a "rainbow government". Ministers of state were blacks, whites, Indians, Coloureds, Muslims, Christians, communists, liberals. When the new Constitution was accepted in 1996, twelve languages were declared official language of the country. In Mandela's solution to the main problems of the world, all cultures meet.

In sum, this book presents some Nobel Laureates and their work, which show us how rich is the human soul and how far is the reach of human empathy.

Anders Hallengren

V. S. NAIPAUL
Photo courtesy of the Nobel Foundation.

Two Worlds[*]

❈

V. S. Naipaul

This is unusual for me. I have given readings and not lectures. I have told people who ask for lectures that I have no lecture to give. And that is true. It might seem strange that a man who has dealt in words and emotions and ideas for nearly fifty years shouldn't have a few to spare, so to speak. But everything of value about me is in my books. Whatever extra there is in me at any given moment isn't fully formed. I am hardly aware of it; it awaits the next book. It will — with luck — come to me during the actual writing, and it will take me by surprise. That element of surprise is what I look for when I am writing. It is my way of judging what I am doing — which is never an easy thing to do.

Proust has written with great penetration of the difference between the writer as writer and the writer as a social being. You will find his thoughts in some of his essays in *Against Sainte-Beuve*, a book reconstituted from his early papers.

The nineteenth-century French critic Sainte-Beuve believed that to understand a writer it was necessary to know as much as possible about

---

[*] Nobel Lecture, December 7, 2001.

the exterior man, the details of his life. It is a beguiling method, using the man to illuminate the work. It might seem unassailable. But Proust is able very convincingly to pick it apart. "This method of Sainte-Beuve," Proust writes, "ignores what a very slight degree of self-acquaintance teaches us: that a book is the product of a different self from the self we manifest in our habits, in our social life, in our vices. If we would try to understand that particular self, it is by searching our own bosoms, and trying to reconstruct it there, that we may arrive at it."

Those words of Proust should be with us whenever we are reading the biography of a writer — or the biography of anyone who depends on what can be called inspiration. All the details of the life and the quirks and the friendships can be laid out for us, but the mystery of the writing will remain. No amount of documentation, however fascinating, can take us there. The biography of a writer — or even the autobiography — will always have this incompleteness.

Proust is a master of happy amplification, and I would like to go back to *Against Sainte-Beuve* just for a little. "In fact," Proust writes, "it is the secretions of one's innermost self, written in solitude and for oneself alone that one gives to the public. What one bestows on private life — in conversation ... or in those drawing-room essays that are scarcely more than conversation in print — is the product of a quite superficial self, not of the innermost self which one can only recover by putting aside the world and the self that frequents the world."

When he wrote that, Proust had not yet found the subject that was to lead him to the happiness of his great literary labour. And you can tell from what I have quoted that he was a man trusting to his intuition and waiting for luck. I have quoted these words before in other places. The reason is that they define how I have gone about my business. I have trusted to intuition. I did it at the beginning. I do it even now. I have no idea how things might turn out, where in my writing I might go next. I have trusted to my intuition to find the subjects, and I have written intuitively. I have an idea when I start, I have a shape; but I will fully understand what I have written only after some years.

I said earlier that everything of value about me is in my books. I will go further now. I will say I am the sum of my books. Each book, intuitively sensed and, in the case of fiction, intuitively worked out, stands on what has gone before, and grows out of it. I feel that at any stage of my literary career it could have been said that the last book contained all the others.

It's been like this because of my background. My background is at once exceedingly simple and exceedingly confused. I was born in Trinidad. It is a small island in the mouth of the great Orinoco river of Venezuela. So Trinidad is not strictly of South America, and not strictly of the Caribbean. It was developed as a New World plantation colony, and when I was born in 1932 it had a population of about 400,000. Of this, about 150,000 were Indians, Hindus and Muslims, nearly all of peasant origin, and nearly all from the Gangetic plain.

This was my very small community. The bulk of this migration from India occurred after 1880. The deal was like this. People indentured themselves for five years to serve on the estates. At the end of this time they were given a small piece of land, perhaps five acres, or a passage back to India. In 1917, because of agitation by Gandhi and others, the indenture system was abolished. And perhaps because of this, or for some other reason, the pledge of land or repatriation was dishonoured for many of the later arrivals. These people were absolutely destitute. They slept in the streets of Port of Spain, the capital. When I was a child I saw them. I suppose I didn't know they were destitute — I suppose that idea came much later — and they made no impression on me. This was part of the cruelty of the plantation colony.

I was born in a small country town called Chaguanas, two or three miles inland from the Gulf of Paria. Chaguanas was a strange name, in spelling and pronunciation, and many of the Indian people — they were in the majority in the area — preferred to call it by the Indian caste name of Chauhan.

I was thirty-four when I found out about the name of my birthplace. I was living in London, had been living in England for sixteen years. I was writing my ninth book. This was a history of Trinidad, a human history, trying to re-create people and their stories. I used to go to the British Museum to read the Spanish documents about the region. These documents — recovered from the Spanish archives — were copied out for the British government in the 1890s at the time of a nasty boundary dispute with Venezuela. The documents begin in 1530 and end with the disappearance of the Spanish Empire.

I was reading about the foolish search for El Dorado, and the murderous interloping of the English hero, Sir Walter Raleigh. In 1595 he raided Trinidad, killed all the Spaniards he could, and went up the Orinoco looking for El Dorado. He found nothing, but when he went back to England he said he had. He had a piece of gold and some sand to show. He said he had hacked the gold out of a cliff on the bank of the Orinoco. The Royal Mint said that the sand he asked them to assay was worthless, and other people said that he had bought the gold beforehand from North Africa. He then published a book to prove his point, and for four centuries people have believed that Raleigh had found something. The magic of Raleigh's book, which is really quite difficult to read, lay in its very long title: *The Discovery of the Large, Rich, and Beautiful Empire of Guiana, with a relation of the great and golden city of Manoa (which the Spaniards call El Dorado) and the provinces of Emeria, Aromaia, Amapaia, and other countries, with their rivers adjoining.* How real it sounds! And he had hardly been on the main Orinoco.

And then, as sometimes happens with confidence men, Raleigh was caught by his own fantasies. Twenty-one years later, old and ill, he was let out of his London prison to go to Guyana and find the gold mines he said he had found. In this fraudulent venture his son died. The father, for the sake of his reputation, for the sake of his lies, had sent his son to his death. And then Raleigh, full of grief, with nothing left to live for, went back to London to be executed.

The story should have ended there. But Spanish memories were long — no doubt because their imperial correspondence was so slow: it might take up to two years for a letter from Trinidad to be read in Spain. Eight years afterwards the Spaniards of Trinidad and Guiana were still settling their scores with the Gulf Indians. One day in the British Museum I read a letter from the King of Spain to the governor of Trinidad. It was dated 12 October 1625.

"I asked you," the King wrote, "to give me some information about a certain nation of Indians called Chaguanes, who you say number above one thousand, and are of such bad disposition that it was they who led the English when they captured the town. Their crime hasn't been punished because forces were not available for this purpose and because the Indians acknowledge no master save their own will. You have decided to give them a punishment. Follow the rules I have given you; and let me know how you get on."

What the governor did I don't know. I could find no further reference to the Chaguanes in the documents in the Museum. Perhaps there were other documents about the Chaguanes in the mountain of paper in the Spanish archives in Seville which the British government scholars missed or didn't think important enough to copy out. What is true is that the little tribe of over a thousand — who would have been living on both sides of the Gulf of Paria — disappeared so completely that no one in the town of Chaguanas or Chauhan knew anything about them. And the thought came to me in the Museum that I was the first person since 1625 to whom that letter of the king of Spain had a real meaning. And that letter had been dug out of the archives only in 1896 or 1897. A disappearance, and then the silence of centuries.

We lived on the Chaguanes' land. Every day in term time — I was just beginning to go to school — I walked from my grandmother's house — past the two or three main-road stores, the Chinese parlour, the Jubilee Theatre, and the high-smelling little Portuguese factory that made cheap blue soap and cheap yellow soap in long bars that were put out to dry and harden in the mornings — every day I walked past these eternal-seeming things — to the Chaguanas Government School. Beyond the school was sugar-cane, estate land, going up to the Gulf of Paria. The people who had been dispossessed would have had their own kind of agriculture, their own calendar, their own codes, their own sacred sites. They would have understood the Orinoco-fed currents in the Gulf of Paria. Now all their skills and everything else about them had been obliterated.

The world is always in movement. People have everywhere at some time been dispossessed. I suppose I was shocked by this discovery in 1967 about my birthplace because I had never had any idea about it. But that was the way most of us lived in the agricultural colony, blindly. There was no plot by the authorities to keep us in our darkness. I think it was more simply that the knowledge wasn't there. The kind of knowledge about the Chaguanes would not have been considered important, and it would not have been easy to recover. They were a small tribe, and they were aboriginal. Such people — on the mainland, in what was called B.G., British Guiana — were known to us, and were a kind of joke. People who were loud and ill-behaved were known, to all groups in Trinidad, I think, as warrahoons. I used to think it was a made-up word, made up to suggest wildness. It was only when I began to travel in Venezuela, in my forties, that I understood that a word like that was the name of a rather large aborginal tribe there.

There was a vague story when I was a child — and to me now it is an unbearably affecting story — that at certain times aboriginal people came across in canoes from the mainland, walked through the forest in the south of the island, and at a certain spot picked some kind of fruit or made some kind of offering, and then went back across the Gulf of Paria to the sodden estuary of the Orinoco. The rite must have been of enormous importance to have survived the upheavals of four hundred years, and the extinction of the aborigines in Trinidad. Or perhaps — though Trinidad and Venezuela have a common flora — they had come only to pick a particular kind of fruit. I don't know. I can't remember anyone inquiring. And now the memory is all lost; and that sacred site, if it existed, has become common ground.

What was past was past. I suppose that was the general attitude. And we Indians, immigrants from India, had that attitude to the island. We lived for the most part ritualised lives, and were not yet capable of self-assessment, which is where learning begins. Half of us on this land of the Chaguanes were pretending — perhaps not pretending, perhaps only feeling, never formulating it as an idea — that we had brought a kind of India with us, which we could, as it were, unroll like a carpet on the flat land.

My grandmother's house in Chaguanas was in two parts. The front part, of bricks and plaster, was painted white. It was like a kind of Indian house, with a grand balustraded terrace on the upper floor, and a prayer-room on the floor above that. It was ambitious in its decorative detail, with lotus capitals on pillars, and sculptures of Hindu deities, all done by people working only from a memory of things in India. In Trinidad it was an architectural oddity. At the back of this house, and joined to it by an upper bridge room, was a timber building in the French Caribbean style. The entrance gate was at the side, between the two houses. It was a tall gate of corrugated iron on a wooden frame. It made for a fierce kind of privacy.

So as a child I had this sense of two worlds, the world outside that tall corrugated-iron gate, and the world at home — or, at any rate, the world of my grandmother's house. It was a remnant of our caste sense, the thing that excluded and shut out. In Trinidad, where as new arrivals we were a disadvantaged community, that excluding idea was a kind of protection; it enabled us — for the time being, and only for the time being — to live in our own way and according to our own rules, to live in our own fading India. It made for an extraordinary self-centredness. We looked inwards; we lived out our days; the world outside existed in a kind of darkness; we inquired about nothing.

There was a Muslim shop next door. The little loggia of my grandmother's shop ended against his blank wall. The man's name was Mian. That was all that we knew of him and his family. I suppose we must have seen him, but I have no mental picture of him now. We knew nothing of Muslims. This idea of strangeness, of the thing to be kept outside, extended even to other Hindus. For example, we ate rice in the middle of the day, and wheat in the evenings. There were some extraordinary people who reversed this natural order and ate rice in the evenings. I thought of these people as strangers — you must imagine me at this time as under seven, because when I was seven all this life of my grandmother's house in Chaguanas came to an end for me. We moved to the capital, and then to the hills to the northwest.

But the habits of mind engendered by this shut-in and shutting-out life lingered for quite a while. If it were not for the short stories my father wrote I would have known almost nothing about the general life of our Indian community. Those stories gave me more than knowledge. They gave me a kind of solidity. They gave me something to stand on in the world. I cannot imagine what my mental picture would have been without those stories.

The world outside existed in a kind of darkness; and we inquired about nothing. I was just old enough to have some idea of the Indian epics, the *Ramayana* in particular. The children who came five years or so after me in our extended family didn't have this luck. No one taught us Hindi. Sometimes someone wrote out the alphabet for us to learn, and that was that; we were expected to do the rest ourselves. So, as English penetrated, we began to lose our language. My grandmother's house was full of religion; there were many ceremonies and readings, some of which went on for days. But no one explained or translated for us who could no longer follow the language. So our ancestral faith receded, became mysterious, not pertinent to our day-to-day life.

We made no inquiries about India or about the families people had left behind. When our ways of thinking had changed, and we wished to know, it was too late. I know nothing of the people on my father's side; I know only that some of them came from Nepal. Two years ago a kind Nepalese who liked my name sent me a copy of some pages from an 1872 gazetteer-like British work about India, *Hindu Castes and Tribes as Represented in Benares*, the pages listed — among a multitude of names — those groups of Nepalese in the holy city of Banaras who carried the name Naipal. That is all that I have.

Away from this world of my grandmother's house, where we ate rice in the middle of the day and wheat in the evenings, there was the great unknown — in this island of only 400,000 people. There were the African or African-derived people who were the majority. They were policemen; they were teachers. One of them was my very first teacher at the Chaguanas Government School; I remembered her with adoration for years. There was the capital, where very soon we would all have to go for education and jobs, and where we would settle permanently, among strangers. There were the white people, not all of them English; and the Portuguese and the Chinese, at one time also immigrants like us. And, more mysterious than these, were the people we called Spanish, *pagnols*, mixed people of warm brown complexions who came from the Spanish time, before the island was detached from Venezuela and the Spanish Empire — a kind of history absolutely beyond my child's comprehension.

To give you this idea of my background, I have had to call on knowledge and ideas that came to me much later, principally from my writing. As a child I knew almost nothing, nothing beyond what I had picked up in my grandmother's house. All children, I suppose, come into the world like that, not knowing who they are. But for the French child, say, that knowledge is waiting. That knowledge will be all around them. It will come indirectly from the conversation of their elders. It will be in the newspapers and on the radio. And at school the work of generations of scholars, scaled down for school texts, will provide some idea of France and the French.

In Trinidad, bright boy though I was, I was surrounded by areas of darkness. School elucidated nothing for me. I was crammed with facts and formulas. Everything had to be learned by heart; everything was abstract for me. Again, I do not believe there was a plan or plot to make our courses like that. What we were getting was standard school learning. In another setting it would have made sense. And at least some of the failing would have lain in me. With my limited social background it was hard for me imaginatively to enter into other societies or societies that were far away. I loved the idea of books, but I found it hard to read them. I got on best with things like Andersen and Aesop, timeless, placeless, not excluding. And when at last in the sixth form, the highest form in the college, I got to like some of our literature texts — Molière, Cyrano de Bergerac — I suppose it was because they had the quality of the fairytale.

When I became a writer those areas of darkness around me as a child became my subjects. The land; the aborigines; the New World; the colony; the history; India; the Muslim world, to which I also felt myself related; Africa; and then England, where I was doing my writing. That was what I meant when I said that my books stand one on the other, and that I am the sum of my books. That was what I meant when I said that my background, the source and prompting of my work, was at once exceedingly simple and exceedingly complicated. You will have seen how simple it was in the country town of Chaguanas. And I think you will understand how complicated it was for me as a writer. Especially in the beginning, when the literary models I had — the models given me by what I can only call my false learning — dealt with entirely different societies. But perhaps you might feel that the material was so rich it would have been no trouble at all to get started and to go on. What I have said about the background, however, comes from the knowledge I acquired with my writing. And you must believe me when I tell you that the pattern in my work has only become clear in the last two months or so. Passages from old books were read to me, and I saw the connections. Until then the greatest trouble for me was to describe my writing to people, to say what I had done.

I said I was an intuitive writer. That was so, and that remains so now, when I am nearly at the end. I never had a plan. I followed no system. I worked intuitively. My aim every time was do a book, to create something that would be easy and interesting to read. At every stage I could only work within my knowledge and sensibility and talent and world-view. Those things developed book by book. And I had to do the books I did because there were no books about those subjects to give me what I wanted. I had to clear up my world, elucidate it, for myself.

I had to go to the documents in the British Museum and elsewhere, to get the true feel of the history of the colony. I had to travel to India because there was no one to tell me what the India my grandparents had come from was like. There was the writing of Nehru and Gandhi; and strangely it was Gandhi, with his South African experience, who gave me more, but not enough. There was Kipling; there were British-Indian writers like John Masters (going very strong in the 1950s, with an announced plan, later abandoned, I fear, for thirty-five connected novels about British India); there were romances by women writers. The few Indian writers who had come up at that time were middle-class people, town-dwellers; they didn't know the India we had come from.

And when that Indian need was satisfied, others became apparent: Africa, South America, the Muslim world. The aim has always been to fill out my world picture, and the purpose comes from my childhood: to make me more at ease with myself. Kind people have sometimes written asking me to go and write about Germany, say, or China. But there is much good writing already about those places; I am willing to depend there on the writing that exists. And those subjects are for other people. Those were not the areas of darkness I felt about me as a child. So, just as there is a development in my work, a development in narrative skill and knowledge and sensibility, so there is a kind of unity, a focus, though I might appear to be going in many directions.

When I began I had no idea of the way ahead. I wished only to do a book. I was trying to write in England, where I stayed on after my years at the university, and it seemed to me that my experience was very thin, was not truly of the stuff of books. I could find in no book anything that came near my background. The young French or English person who wished to write would have found any number of models to set him on his way. I had none. My father's stories about our Indian community belonged to the past. My world was quite different. It was more urban, more mixed. The simple physical details of the chaotic life of our extended family — sleeping rooms or sleeping spaces, eating times, the sheer number of people — seemed impossible to handle. There was too much to be explained, both about my home life and about the world outside. And at the same time there was also too much about us — like our own ancestry and history — that I didn't know.

At last one day there came to me the idea of starting with the Port of Spain street to which we had moved from Chaguanas. There was no big corrugated-iron gate shutting out the world there. The life of the street was open to me. It was an intense pleasure for me to observe it from the verandah. This street life was what I began to write about. I wished to write fast, to avoid too much self-questioning, and so I simplified. I suppressed the child-narrator's background. I ignored the racial and social complexities of the street. I explained nothing. I stayed at ground level, so to speak. I presented people only as they appeared on the street. I wrote a story a day. The first stories were very short. I was worried about the material lasting long enough. But then the writing did its magic. The material began to present itself to me from many sources. The stories became longer; they couldn't be written in a day. And then the inspiration, which at one stage had seemed very easy, rolling me along, came to an end. But a book had been written, and I had in my own mind become a writer.

The distance between the writer and his material grew with the two later books; the vision was wider. And then intuition led me to a large book about our family life. During this book my writing ambition grew. But when it was over I felt I had done all that I could do with my island material. No matter how much I meditated on it, no further fiction would come.

Accident, then, rescued me. I became a traveller. I travelled in the Caribbean region and understood much more about the colonial set-up of which I had been part. I went to India, my ancestral land, for a year; it was a journey that broke my life in two. The books that I wrote about these two journeys took me to new realms of emotion, gave me a world-view I had never had, extended me technically. I was able in the fiction that then came to me to take in England as well as the Caribbean — and how hard that was to do. I was able also to take in all the racial groups of the island, which I had never before been able to do.

This new fiction was about colonial shame and fantasy, a book, in fact, about how the powerless lie about themselves, and lie to themselves, since it is their only resource. The book was called *The Mimic Men*. And it was not about mimics. It was about colonial men mimicking the condition of manhood, men who had grown to distrust everything about themselves. Some pages of this book were read to me the other day — I hadn't looked at it for more than thirty years — and it occurred to me that I had been writing about colonial schizophrenia. But I hadn't thought of it like that. I had never used abstract words to describe any writing purpose of mine. If I had, I would never have been able to do the book. The book was done intuitively, and only out of close observation.

I have done this little survey of the early part of my career to try to show the stages by which, in just ten years, my birthplace had altered or developed in my writing: from the comedy of street life to a study of a kind of widespread schizophrenia. What was simple had become complicated.

Both fiction and the travel-book form have given me my way of looking; and you will understand why for me all literary forms are equally valuable. It came to me, for instance, when I set out to write my third book about India — twenty-six years after the first — that what was most important about a travel book were the people the writer travelled among. The people had to define themselves. A simple enough idea, but it required a new kind of book; it called for a new way of travelling. And it was the very method I used later when I went, for the second time, into the Muslim world.

I have always moved by intuition alone. I have no system, literary or political. I have no guiding political idea. I think that probably lies with my ancestry. The Indian writer R. K. Narayan, who died this year, had no political idea. My father, who wrote his stories in a very dark time, and for no reward, had no political idea. Perhaps it is because we have been far from authority for many centuries. It gives us a special point of view. I feel we are more inclined to see the humour and pity of things.

Nearly thirty years ago I went to Argentina. It was at the time of the guerrilla crisis. People were waiting for the old dictator Perón to come back from exile. The country was full of hate. Peronists were waiting to settle old scores. One such man said to me, "There is good torture and bad torture." Good torture was what you did to the enemies of the people. Bad torture was what the enemies of the people did to you. People on the other side were saying the same thing. There was no true debate about anything. There was only passion and the borrowed political jargon of Europe. I wrote, "Where jargon turns living issues into abstractions, and where jargon ends by competing with jargon, people don't have causes. They only have enemies."

And the passions of Argentina are still working themselves out, still defeating reason and consuming lives. No resolution is in sight.

I am near the end of my work now. I am glad to have done what I have done, glad creatively to have pushed myself as far as I could go. Because of the intuitive way in which I have written, and also because of the baffling nature of my material, every book has come as a blessing. Every book has amazed me; up to the moment of writing I never knew it was there. But the greatest miracle for me was getting started. I feel — and the anxiety is still vivid to me — that I might easily have failed before I began.

I will end as I began, with one of the marvellous little essays of Proust in *Against Sainte-Beuve*. "The beautiful things we shall write if we have talent," Proust says, "are inside us, indistinct, like the memory of a melody which delights us though we are unable to recapture its outline. Those who are obsessed by this blurred memory of truths they have never known are the men who are gifted ... Talent is like a sort of memory which will enable them finally to bring this indistinct music closer to them, to hear it clearly, to note it down ..."

Talent, Proust says. I would say luck, and much labour.

❋

Sir V. S. Naipaul receiving his Nobel Prize from His Majesty the King at the Stockholm Concert Hall.

Photo courtesy of Hans Mehlin and the Nobel Foundation.

The Enigma of Arrival[*]

❋

V. S. Naipaul

To see the possibility, the certainty, of ruin, even at the moment of creation: it was my temperament. Those nerves had been given me as a child in Trinidad partly by our family circumstances: the half-ruined or broken-down houses we lived in, our many moves, our general uncertainty. Possibly, too, this mode of feeling went deeper, and was an ancestral inheritance, something that came with the history that had made me: not only India, with its ideas of a world outside men's control, but also the colonial plantations or estates of Trinidad, to which my impoverished Indian ancestors had been transported in the last century — estates of which this Wiltshire estate, where I now lived, had been the apotheosis.

Fifty years ago there would have been no room for me on the estate; even now my presence was a little unlikely. But more than accident had brought me here. Or rather, in the series of accidents that had brought me to the manor cottage, with a view of the restored church, there was a clear historical line. The migration, within the British Empire, from India to

---

[*] Excerpt from the novel *The Enigma of Arrival* (Chapter: Jack's Garden, pp. 52–53) © V. S. Naipaul, 1987.

Trinidad had given me the English language as my own, and a particular kind of education. This had partly seeded my wish to be a writer in a particular mode, and had committed me to the literary career I had been following in England for twenty years.

The history I carried with me, together with the self-awareness that had come with my education and ambition, had sent me into the world with a sense of glory dead; and in England had given me the rawest stranger's nerves. Now ironically — or aptly — living in the grounds of . this shrunken estate, going out for my walks, those nerves were soothed, and in the wild garden and orchard beside the water meadows I found a physical beauty perfectly suited to my temperament and answering, besides, every good idea I could have had, as a child in Trinidad, of the physical aspect of England.

The estate had been enormous, I was told. It had been created in part by the wealth of empire. But then bit by bit it had been alienated. The family in its many branches flourished in other places. Here in the valley there now lived only my landlord, elderly, a bachelor, with people to look after him. Certain physical disabilities had now been added to the malaise which had befallen him years before, a malaise of which I had no precise knowledge, but interpreted as something like accidia, the monk's torpor or disease of the Middle Ages — which was how his great security, his excessive worldly blessings, had taken him. The accidia had turned him into a recluse, accessible only to his intimate friends. So that on the manor itself, as on my walks on the down, I had a kind of solitude.

I felt a great sympathy for my landlord. I felt I could understand his malaise; I saw it as the other side of my own. I did not think of my landlord as a failure. Words like failure and success didn't apply. Only a grand man or a man with a grand idea of his human worth could ignore the high money value of his estate and be content to live in its semi-ruin. My meditations in the manor were not of imperial decline. Rather, I wondered at the historical chain that had brought us together — he in his house, I in his cottage, the wild garden his taste (as I was told) and also mine.

❊

Naipaul with his cat.
© *Jerry Bauer*

Nadine GORDIMER

Photo courtesy of the Nobel Foundation.

Writing and Being*

Nadine Gordimer

In the beginning was the Word.

The Word was with God, signified God's Word, the word that was Creation. But over the centuries of human culture the word has taken on other meanings, secular as well as religious. To have the word has come to be synonymous with ultimate authority, with prestige, with awesome, sometimes dangerous persuasion, to have Prime Time, a TV talk show, to have the gift of the gab as well as that of speaking in tongues. The word flies through space, it is bounced from satellites, now nearer than it has ever been to the heaven from which it was believed to have come. But its most significant transformation occured for me and my kind long ago, when it was first scratched on a stone tablet or traced on papyrus, when it materialized from sound to spectacle, from being heard to being read as a series of signs, and then a script; and travelled through time from parchment to Gutenberg. For this is the genesis story of the writer. It is the story that wrote her or him into being.

---

* Nobel Lecture, December 7, 1991.

It was, strangely, a double process, creating at the same time both the writer and the very purpose of the writer as a mutation in the agency of human culture. It was both ontogenesis as the origin and development of an individual being, and the adaptation, in the nature of that individual, specifically to the exploration of ontogenesis, the origin and development of the individual being. For we writers are evolved for that task. Like the prisoners incarcerated with the jaguar in Borges' story[1], 'The God's Script', who was trying to read, in a ray of light which fell only once a day, the meaning of being from the marking on the creature's pelt, we spend our lives attempting to interpret through the word the readings we take in the societies, the world of which we are part. It is in this sense, this inextricable, ineffable participation, that writing is always and at once an exploration of self and of the world; of individual and collective being.

Being here.

Humans, the only self-regarding animals, blessed or cursed with this torturing higher faculty, have always wanted to know why. And this is not just the great ontological question of why we are here at all, for which religions and philosophies have tried to answer conclusively for various peoples at various times, and science tentatively attempts dazzling bits of explantation we are perhaps going to die out in our millenia, like dinosaurs, without having developed the necessary comprehension to understand as a whole. Since humans became self-regarding they have sought, as well, explanations for the common phenomena of procreation, death, the cycle of seasons, the earth, sea, wind and stars, sun and moon, plenty and disaster. With myth, the writer's ancestors, the oral story-tellers, began to feel out and formulate these mysteries, using the elements of daily life — observable reality — and the faculty of the imagination — the power of projection into the hidden — to make stories.

Roland Barthes[2] asks, 'What is characteristic of myth?' And answers: 'To transform a meaning into form.' Myths are stories that mediate in this way between the known and unknown. Claude Lévi-Strauss[3] wittily de-mythologizes myth as a genre between a fairy tale and a detective story. Being here; we don't know who-dun-it. But something satisfying, if not the answer, can be invented. Myth was the mystery plus the fantasy — gods, anthropomorphized animals and birds, chimera, phantasmagorical creatures — that posits out of the imagination some sort of explanation for the mystery. Humans and their fellow creatures were the materiality of the story, but as Nikos Kazantzakis[4] once wrote, 'Art is the representation not of the body but of the forces which created the body.'

There are many proven explanations for natural phenomena now; and there are new questions of being arising out of some of the answers. For this reason, the genre of myth has never been entirely abandoned, although we are inclined to think of it as archaic. If it dwindled to the children's bedtime tale in some societies, in parts of the world protected by forests or deserts from international megaculture it has continued, alive, to offer art as a system of mediation between the individual and being. And it has made a whirling comeback out of Space, an Icarus in the avatar of Batman and his kind, who never fall into the ocean of failure to deal with the gravity forces of life. These new myths, however, do not seek so much to enlighten and provide some sort of answers as to distract, to provide a fantasy escape route for people who no longer want to face even the hazard of answers to the terrors of their existence. (Perhaps it is the positive knowledge that humans now possess the means to destroy their whole planet, the fear that they have in this way themselves become the gods, dreadfully charged with their own continued existence, that has made comic-book and movie myth escapist.) The forces of being remain. They are what the writer, as distinct from the contemporary popular mythmaker, still engage today, as myth in its ancient form attempted to do.

How writers have approached this engagement and continue to experiment with it has been and is, perhaps more than ever, the study of literary scholars. The writer in relation to the nature of perceivable reality and what is beyond — imperceivable reality — is the basis for all these studies, no matter what resulting concepts are labelled, and no matter in what categorized microfiles writers are stowed away for the annals of literary historiography. Reality is constructed out of many elements and entities, seen and unseen, expressed, and left unexpressed for breathing-space in the mind. Yet from what is regarded as old-hat psychological analysis to modernism and post-modernism, structuralism and poststructuralism, all literary studies are aimed at the same end: to pin down to a consistency (and what is consistency if not the principle hidden within the riddle?); to make definitive through methodology the writer's grasp at the forces of being. But life is aleatory in itself; being is constantly pulled and shaped this way and that by circumstances and different levels of consciousness. There is no pure state of being, and it follows that there is no pure text, 'real' text, totally incorporating the aleatory. It surely cannot be reached by any critical methodology, however interesting the attempt. To deconstruct a text is in a way a contradiction, since to deconstruct it is to make another construction out of the pieces, as Roland Barthes[5] does so fascinatingly, and admits to, in his linguistic and semantical dissection of Balzac's story, 'Sarrasine'. So the literary scholars end up being some kind of storyteller, too.

Perhaps there is no other way of reaching some understanding of being than through art? Writers themselves don't analyze what they do; to analyze would be to look down while crossing a canyon on a tightrope. To say this is not to mystify the process of writing but to make an image out of the intense inner concentration the writer must have to cross the chasms of the aleatory and make them the word's own, as an explorer plants a flag. Yeats' inner 'lonely impulse of delight' in the pilot's solitary flight, and his 'terrible beauty' born of mass uprising, both opposed and conjoined; E. M. Forster's modest 'only connect'; Joyce's chosen, wily 'silence, cunning and exile'; more contemporary, Gabriel García Márquez's labyrinth in which power over others, in the person of Simón Bolívar, is led to the thrall of the only unassailable power, death — these are some examples of the writer's endlessly varied ways of approaching the state of being through the word. Any writer of any worth at all hopes to play only a pocket-torch of light — and rarely, through genius, a sudden flambeau — into the bloody yet beautiful labyrinth of human experience, of being.

Anthony Burgess[6] once gave a summary definition of literature as 'the aesthetic exploration of the world'. I would say that writing only begins there, for the exploration of much beyond, which nevertheless only aesthetic means can express.

How does the writer become one, having been given the word? I do not know if my own beginnings have any particular interest. No doubt they have much in common with those of others, have been described too often before as a result of this yearly assembly before which a writer stands. For myself, I have said that nothing factual that I write or say will be as truthful as my fiction. The life, the opinions, are not the work, for it is in the tension between standing apart and being involved that the imagination transforms both. Let me give some minimal account of myself. I am what I suppose would be called a natural writer. I did not make any decision to become one. I did not, at the beginning, expect to earn a living by being read. I wrote as a child out of the joy of apprehending life through my senses — the look and scent and feel of things; and soon out of the emotions that puzzled me or raged within me and which took form, found some enlightenment, solace and delight, shaped in the written word. There is a little Kafka[7] parable that goes like this; 'I have three dogs: Hold-him, Seize-him, and Nevermore. Hold-him and Seize-him are ordinary little Schipperkes and nobody would notice them if they were alone. But there is Nevermore, too. Nevermore is a mongrel Great Dane and has an appearance that centuries of the most careful breeding could never have produced. Nevermore is a gypsy.' In the small South African gold-mining town where I was growing up I was Nevermore the mongrel (although I could scarely have been described as a Great Dane ...) in whom the accepted characteristics of the townspeople could not be traced. I was the Gypsy, tinkering with words second-hand, mending my own efforts at writing by learning from what I read. For my school was the local library. Proust, Chekhov and Dostoevsky, to name only a few to whom I owe my existence as a writer, were my professors. In that period of my life, yes, I was evidence of the theory that books are made out of other books ... But I did not remain so for long, nor do I believe any potential writer could.

With adolescence comes the first reaching out to otherness through the drive of sexuality. For most children, from then on the faculty of the imagination, manifest in play, is lost in the focus on day dreams of desire and love, but for those who are going to be artists of one kind or another the first life-crisis after that of birth does something else in addition: the imagination gains range and extends by the subjective flex of new and turbulent emotions. There are new perceptions. The writer begins to be able to enter into other lives. The process of standing apart and being involved has come.

Unknowingly, I had been addressing myself on the subject of being, whether, as in my first stories, there was a child's contemplation of death and murder in the necessity to finish off, with a death blow, a dove mauled by a cat, or whether there was wondering dismay and early consciousness of racism that came of my walk to school, when on the way I passed storekeepers, themselves East European immigrants kept lowest in the ranks of the Anglo-Colonial social scale for whites in the mining town, roughly those whom colonial society ranked lowest of all, discounted as less than human — the black miners who were the stores' customers. Only many years later was I to realize that if I had been a child in that category — black — I might not have become a writer at all, since the library that made this possible for me was not open to any black child. For my formal schooling was sketchy, at best.

To adress oneself to others begins a writer's next stage of development. To publish to anyone who would read what I wrote. That was my natural, innocent assumption of what publication meant, and it has not changed, that is what it means to me today, in spite of my awareness that most people refuse to believe that a writer does not have a particular audience in mind; and my other awareness: of the temptations, conscious and unconscious, which lure the writer into keeping a corner of the eye on who will take offense, who will approve what is on the page — a temptation that, like Eurydice's straying glance, will lead the writer back into the Shades of a destroyed talent.

The alternative is not the malediction of the ivory tower, another destroyer of creativity. Borges once said he wrote for his friends and to pass the time. I think this was an irritated flippant response to the crass question — often an accusation — 'For whom do you write?', just as Sartre's admonition that there are times when a writer should cease to write, and act upon being only in another way, was given in the frustration of an unresolved conflict between distress at injustice in the world and the knowledge that what he knew how to do best was write.

Both Borges and Sartre, from their totally different extremes of denying literature a social purpose, were certainly perfectly aware that it has its implicit and unalterable social role in exploring the state of being, from which all other roles, personal among friends, public at the protest demonstration, derive. Borges was not writing for his friends, for he published and we all have received the bounty of his work. Sartre did not stop writing, although he stood at the barricades in 1968.

The question of for whom do we write nevertheless plagues the writer, a tin can attached to the tail of every work published. Principally it jangles the inference of tendentiousness as praise or denigration. In this context, Camus[8] dealt with the question best. He said that he liked individuals who take sides more than literatures that do. 'One either serves the whole of man or does not serve him at all. And if man needs bread and justice, and if what has to be done must be done to serve this need, he also needs pure beauty which is the bread of his heart.' So Camus called for 'Courage in and talent in one's work.' And Márquez[9] redefined tender fiction thus: The best way a writer can serve a revolution is to write as well as he can.

I believe that these two statements might be the credo for all of us who write. They do not resolve the conflicts that have come, and will continue to come, to contemporary writers. But they state plainly an honest possibility of doing so, they turn the face of the writer squarely to her and his existence, the reason to be, as a writer, and the reason to be, as a responsible human, acting, like any other, within a social context.

Being here: in a particular time and place. That is the existential position with particular implications for literature. Czeslaw Milosz[10] once wrote the cry: 'What is poetry which does not serve nations or people?' and Brecht[11] wrote of a time when 'to speak of trees is almost a crime'. Many of us have had such despairing thoughts while living and writing through such times, in such places, and Sartre's solution makes no sense in a world where writers were — and still are — censored and forbidden to write, where, far from abandoning the word, lives were and are at risk in smuggling it, on scraps of paper, out of prisons. The state of being whose ontogenesis we explore has overwhelmingly included such experiences. Our approaches, in Nikos Kazantzakis'[12] words, have to 'make the decision which harmonizes with the fearsome rhythm of our time.'

Some of us have seen our books lie for years unread in our own countries, banned, and we have gone on writing. Many writers have been imprisoned. Looking at Africa alone — Soyinka, Ngugi wa Thiong'o, Jack Mapanje, in their countries, and in my own country, South Africa, Jeremy Cronin, Mongane Wally Serote, Breyten Breytenbach, Dennis Brutus, Jaki Seroke: all these went to prison for the courage shown in their lives, and have continued to take the right, as poets, to speak of trees. Many of the greats, from Thomas Mann to Chinua Achebe, cast out by political conflict and oppression in different countries, have endured the trauma of exile, from which some never recover as writers, and some do not survive at all. I think of the South Africans, Can Themba, Alex la Guma, Nat Nakasa, Todd Matshikiza. And some writers, over half a century from Joseph Roth to Milan Kundera, have had to publish new works first in the word that is not their own, a foreign language.

Then in 1988 the fearsome rhythm of our time quickened in an unprecedented frenzy to which the writer was summoned to submit the word. In the broad span of modern times since the Enlightenment writers have suffered opprobrium, bannings and even exile for other than political reasons. Flaubert dragged into court for indecency, over *Madame Bovary*, Strindberg arraigned for blasphemy, over *Marrying*, Lawrence's *Lady Chatterley's Lover* banned — there have been many examples of so-called offense against hypocritical bourgeois mores, just as there have been of treason against political dictatorships. But in a period when it would be unheard of for countries such as France, Sweden and Britain to bring such charges against freedom of expression, there has risen a force that takes its appalling authority from something far more widespread than social mores, and far more powerful than the power of any single political regime. The edict of a world religion has sentenced a writer to death.

For more than three years, now, wherever he is hidden, wherever he might go, Salman Rushdie has existed under the Muslim pronouncement upon him of the fatwa. There is no asylum for him anywhere.

Every morning when this writer sits down to write, he does not know if he will live through the day; he does not know whether the page will ever be filled. Salman Rushdie happens to be a brilliant writer, and the novel for which he is being pilloried, *The Satanic Verses*, is an innovative exploration of one of the most intense experiences of being in our era, the individual personality in transition between two cultures brought together in a post-colonial world. All is re-examined through the refraction of the imagination; the meaning of sexual and filial love, the rituals of social acceptance, the meaning of a formative religious faith for individuals removed from its subjectivity by circumstance opposing different systems of belief, religious and secular, in a different context of living. His novel is a true mythology. But although he has done for the postcolonial consciousness in Europe what Günter Grass did for the post-Nazi one with *The Tin Drum* and *Dog Years*, perhaps even has tried to approach what Beckett did for our existential anguish in *Waiting For Godot*, the level of his achievement should not matter. Even if he were a mediocre writer, his situation is the terrible concern of every fellow writer for, apart from his personal plight, what implications, what new threat against the carrier of the word does it bring? It should be the concern of individuals and above all, of governments and human rights organizations all over the world. With dictatorships apparently vanquished, this murderous new dictate invoking the power of international terrorism in the name of a great and respected religion should and can be dealt with only by democratic governments and the United Nations as an offense against humanity.

I return from the horrific singular threat to those that have been general for writers of this century now in its final, summing-up decade. In repressive regimes anywhere — whether in what was the Soviet bloc, Latin America, Africa, China — most imprisoned writers have been shut away for their activities as citizens striving for liberation against the oppression of the general society to which they belong. Others have been condemned by repressive regimes for serving society by writing as well as they can; for this aesthetic venture of ours becomes subversive when the shameful secrets of our times are explored deeply, with the artist's rebellious-integrity to the state of being manifest in life around her or him; then the writer's themes and characters inevitably are formed by the pressures and distortions of that society as the life of the fisherman is determined by the power of the sea.

There is a paradox. In retaining this integrity, the writer sometimes must risk both the state's indictment of treason, and the liberation forces' complaint of lack of blind commitment. As a human being, no writer can stoop to the lie of Manichean 'balance'. The devil always has lead in his shoes, when placed on his side of the scale. Yet, to paraphrase coarsely Márquez's dictum given by him both as a writer and a fighter for justice, the writer must take the right to explore, warts and all, both the enemy and the beloved comrade in arms, since only a try for the truth makes sense of being, only a try for the truth edges towards justice just ahead of Yeats's beast slouching to be born. In literature, from life,

> *we page through each other's faces*
> *we read each looking eye*
> *... It has taken lives to be able to do so.*

These are the words of the South African poet and fighter for justice and peace in our country, Mongane Serote.[13]

The writer is of service to humankind only insofar as the writer uses the word even against his or her own loyalties, trusts the state of being, as it is revealed, to hold somewhere in its complexity filaments of the cord of truth, able to be bound together, here and there, in art: trusts the state of being to yield somewhere fragmentary phrases of truth, which is the final word of words, never changed by our stumbling efforts to spell it out and write it down, never changed by lies, by semantic sophistry, by the dirtying of the word for the purposes of racism, sexism, prejudice, domination, the glorification of destruction, the curses and the praise-songs.

Endnotes

1. "The God's Script" from *Labyrinths: Selected Stories & Other Writings* by Jorge Luis Borges. Edited by Donald A. Yates and James E. Irby. Penguin Modern Classics, 1970, page 71.

2. *Mythologies* by Roland Barthes. Translated by Annette Lavers. Hill & Wang, 1972, page 131.

3. *Histoire de Lynx* by Claude Lévi-Strauss. '… je les situais à mi-chemin entre le conte de fées et le roman policier'. Plon, 1991, page 13.

4. *Report to Greco* by Nikos Kazantzakis. Faber & Faber, 1973, page 150.

5. *S/Z* by Roland Barthes. Translated by Richard Miller. Jonathan Cape, 1975.

6. London *Observer* review. 19/4/81. Anthony Burgess.

7. *Wedding Preparations in the Country* by Franz Kafka. Secker & Warburg, 1973.

8. *Carnets* 1942–5 by Albert Camus.

9. Gabriel García Márquez. In an interview; my notes do not give the journal or date.

10. 'Dedication' from *Selected Poems* by Czeslaw Milosz. Ecco Press, 1980.

11. 'To Posterity' from *Selected Poems* by Bertolt Brecht. Translated by H. R. Hays. Grove Press, 1959, page 173.

12. *Report to Greco* by Nikos Kazantzakis. Faber & Faber.

13. *A Tough Tale* by Mongane Wally Serote. Kliptown Books, 1987.

❄

Per WÄSTBERG, Reinhold CASSIRER and Nadine GORDIMER
Photo courtesy of Anita Theorell and the Nobel Foundation.

Nadine Gordimer and the South African Experience

❄

Per Wästberg

Warrior of the Imagination

Nadine Gordimer, born in 1923 and, in Seamus Heaney's words, one of "the guerrillas of the imagination," became the first South African and the seventh woman to be awarded the Nobel Prize for Literature in 1991.

Over half a century, Gordimer has written thirteen novels, over two hundred short stories, and several volumes of essays. Ten books are devoted to her works, and about two hundred critical essays appear in her bibliography. Few living authors have kept so many academics occupied. The best study, in my view, is Stephen Clingman's *The Novels of Nadine Gordimer: History from the Inside* (London: Bloomsbury, 1993).

Gordimer's works have been translated into more than thirty languages. She herself has been awarded fifteen honorary doctorates and received major literary prizes. And she has given much personal support to individual writers.

Geiger Counter of Apartheid

Through the years, visitors have come to her house to inform, plead, and confess. She has been so deeply involved in the anti-apartheid struggle that one wonders how she managed to keep her integrity and observe South African society with such a discerning eye in her stories. In spite of her taking part in demonstrations, giving speeches, and travelling around the world supporting good causes, Gordimer is intensely private and guards her study, staying there through the mornings up to a late lunch. She does not make friends easily, says her oldest friend Anthony Sampson, but when she does she often retains them for life.

Gordimer endured the bleak decades, refusing to move abroad as so many others did. Her husband, Reinhold Cassirer, was a refugee from Nazi Germany, who served in the British Army in World War II. Her daughter settled in France, her son in New York; but she kept her lines open inside South Africa, out of commitment to black liberation and also for the sake of her own creativity and that of black South African writers who were silenced, for whom she had to speak.

Gordimer's Nobel Prize put the searchlight on a country in painful transition from an oppressive racism to a turbulent democracy. South Africa's literature is rich. But beyond doubt, Nadine Gordimer is the writer that most stubbornly has kept the true face of racism in front of us, in all its human complexities.

For fifty years, Gordimer has been the Geiger counter of apartheid and of the movements of people across the crust of South Africa. Her work reflects the psychic vibrations within that country, the road from passivity and blindness to resistance and struggle, the forbidden friendships, the censored soul, and the underground networks. She has outlined a free zone where it was possible to try out, in imagination, what life beyond apartheid might be like. She wrote as if censorship did not exist and as if there were readers willing to listen. In her characters, the major currents of contemporary history intersect.

Gordimer has created individuals who make their moral choices behind private doors and in the public sphere. She has painted a social background subtler than anything presented by political scientists, thus providing an insight into the roots of the struggle and the mechanisms of change that no historian could have matched.

Nelson Mandela and the African National Congress

Early in her career, before other white writers, Gordimer saw the inventive buoyancy and playful courage of Sophiatown's and Soweto's black intellectuals and politicians, the circles where the young Nelson Mandela moved. Nadine's best friend, Bettie du Toit, was arrested in 1960, the year of Sharpeville, and so the political struggle entered her life. When Mandela and his colleagues were on trial for their lives, she became a close friend of their defence lawyers, Bram Fischer (the subject of *Burger's Daughter*) and George Bizos. Indeed, her proudest day, she says, was not when she was awarded the Nobel Prize (of which she gave a portion to the South African Congress of Writers) but when she testified at the Delmas trial in 1986, to save the lives of twenty-two ANC members, all of them accused of treason.

When Mandela was freed, Nadine Gordimer was one of the first he wished to see. "Strange to live in a country where there are still heroes." (*Burger's Daughter*)

Asked what she would write about when apartheid was over, Gordimer replied, "Life didn't end with apartheid; new life began." With her novels of the mid-1990s, *None to Accompany Me* and *The House Gun*, Gordimer proved that there is literary life after apartheid. In fact, her imagination was unbound; her books catch the social ambiguities of her time. She has not suffered the fate of some of her East European colleagues.

Race and Gender

The writer's task is to transform experience, to enter into the existence of others, whether they be black or white, men or women, and to use the tension in both participating and standing at the side. With her restless energy and prodigious discipline, Gordimer is able to put herself not only in the mind, but also in the body of criminal and saint, male or female, black or white. She herself contains many persons in one body:

she grew up speaking English in the African mining town of Springs, was a Jewish girl in a Catholic convent school, and then was educated at home. Lonely, with a domineering mother, she wrote from an early age and published her first adult story at 15.

Her father was a Jewish watchmaker from the border between Latvia and Lithuania. He opened a jeweller's shop in Springs in Transvaal and sold trophy cups to shooting clubs, as well as engagement rings. He read nothing. Her mother, a transplanted Londoner, read aloud to her daughters. Troubled by the way blacks were treated, she founded a crèche, a nursery school for black children. Gordimer's father, on the other hand, to avoid being conspicuous, turned a blind eye to any reminder of the oppression he had himself been subjected to in czarist Russia. This is the world of *The Lying Days*.

Without the library in her small town, Nadine may not have become a writer; she was well aware that blacks were not permitted to use the local library.

The Novel as History

Her first published volume appeared in 1949, the short-story collection *Face to Face*. *The Lying Days*, published in 1953, is about waking up from the naïveté of a small colonial town. Gordimer wrote of "having a picnic in a beautiful cemetery where people were buried alive." South Africa is seen as a whites-only annex of European society, with middle-class suburbs, Sunday outings, and a blindness about anything lurking below the surface. The vast black population is regarded as if it was there only to serve whites in industry and at home.

"The novel as history is something other than a historical novel," Nadine Gordimer has remarked. Her protagonists and their points of view are constantly shifting. It may be Hillela, the sexual rebel, amoral and intuitive, demolishing apartheid in her personal sphere; or Bray in *A Guest of Honour*, a good man, a fragile liberal, betrayed by the old empire that found him too radical and by the new that tramples him down in passing. Through the novels, Gordimer's historical consciousness grows. In *A World of Strangers* (1958), we find the dilemma of well-meaning liberalism, while in *Occasion for Loving*

(1963), it is the insight of the humanist that apartheid cannot be reformed by pious words. *The Late Bourgeois World* (1966) reflects Nelson Mandela's decision to switch from passive resistance to sabotage.

Gordimer joined the ANC before it was legal to do so and was impatient with whites who accused the ANC of autocratic tendencies instead of influencing it by joining. She has herself been both loyal and critical, all the time safeguarding the integrity of her imagination. Her inspiration, rather than her cause for despair, are the dangerous but rewarding contradictions of South African society today.

Love and Politics

When I first met Nadine Gordimer in early 1959, she had just published *A World of Strangers* and moved from the exploration of her upbringing in *The Lying Days* (1953) to her first attempt to focus on the growing anti-apartheid movement and the multiracialism of the *Drum* set of Can Themba, Lewis Nkosi, and others. Her third novel, *Occasion for Loving* (1963), deals with the failure of tolerance and humanism; the increasing absurdity of the race laws brought friendship and love across the colour bar to a halt. In her fourth novel, *The Late Bourgeois World* (1966), the choice is between the naive idealism of saboteurs or the well-meaning cynicism of passive liberals.

In 1971, Gordimer published *A Guest of Honour*, a huge novel about the birth pangs of the new Africa. Individual history and great ideological perspectives are woven into a chronicle whose protagonists embody the social, political, and moral problems arising when a victorious liberation front splits up into factions. Idealism and good will are almost drowned by a new brutality and a corruption similar to that under colonial rule. It deals with policy formulation and backroom bargaining and uses trade union jargon, local language transposed into English, settler ironies, and nationalist slogans. It is a Henry Jamesian enterprise where society and marriage, politics and landscapes, mix without obscuring the pattern.

The Conservationist, which won the Booker Prize for 1974, evokes the sterility of the white community. This novel is a kind of sequel to the first classic of South African literature, Olive Schreiner's *The Story of an African Farm* (1883), which can also be said of another remarkable novel centered on a farm, J. M. Coetzee's *In the Heart of the Country* (1977). Mehring, the Afrikaner antihero whose farm is as barren as his life, conserves both nature and the apartheid system, the one to keep the other at bay. He likes to preserve nature's variety but is in fact its exploiter; nor does nature return his sentimental love. In his moral vacuum, Mehring sees Africa returning to the possession of the blacks. Gordimer's powerful landscape descriptions become metaphors of the soul. Using Zulu creation myths, she looks in a new way at nature in South Africa, leaving her white predecessors in art and literature behind.

The Conservationist is a novel of ironies. Mehring is not a male chauvinist Boer; he is tolerant but no liberal, a financier using his farm as a tax-deductible expense. His leftist mistress travels round the world on his money. He likes to be seen as a country gentleman, but sexually he is a colonialist as we see when he picks up a coloured girl and takes her to an old mine property, only to be surprised by the mine guards.

The corpse of an unknown African is found on the farm, silently disputing Mehring's claim to his own clean soil. He identifies with the nameless black man under the reeds, burying him in a coffin. Yet the corpse haunting Mehring and his house (a symbol of South Africa) is the claim on Africa by those who possess no land at all.

The Conservationist is Gordimer's densest and most poetical novel. Its minute details and documentary precision form an intricate web of meanings where each stone, egg, and piece of marble carry symbolic implications. Here, as in *July's People*, Gordimer finds a fertile blend of narrative interest, rich language, and high moral seriousness, as well as rounded characters. She avoids explanations and leaves the reader free to interpret.

A Dialogue with the Future

Burger's Daughter (1979) is, in her own words, "a coded homage" to Bram Fischer, the communist lawyer who was sentenced to life in prison and whose name nobody was allowed to mention. Gordimer never claimed to portray him — although his daughter recognised their lives — but to convey the hidden truth behind a public person. The challenge to the writer is to penetrate official lies and facades, to see beyond and behind, with an intuition and insight unhampered by social conventions or family discretion. She intended, she said, "to bring to a broad canvas the position of the white Left in South Africa, and the extraordinary dynasty of belief and struggle in these families."

Two of her major works, *July's People* (1981) and *My Son's Story* (1990), deal, on several symbolic levels, with individual fates and the terrible choices forced upon people by an inhuman ideology. In the latter, Gordimer catches both the unexpected moment when the revolutionary spark ignites and the daily routine when internal dissension rocks the upper reaches of the anti-apartheid movement. The novel's central character, a mixed race man named Sonny, is trapped between being a teacher and a politician, a father and a husband. About to enter a political collective struggle, he is caught between one state and another to come; he is himself the transition. Through him, and others, Gordimer enters into a dialogue with the future, with the absent forces that are to rule our lives in years ahead.

Gordimer reveals situations when reality suddenly takes another course and we are caught in our roles and expectations, in the traps of skin colour, class, family, and the body itself. She is drawn to those who try to escape from the trap: What makes the suburban housewife become an underground agent, the lawyer to sacrifice his life for a future not his, the young architect to hide a black freedom fighter?

How do faithfulness and betrayal interact in an erotic and political context?

July's People are Maureen and Bamford Smales, he an architect, she a housewife and former dancer, with three children, a nice suburban house, and a servant named July. It is a parable of the future: the servant hides his master's family to protect them from catastrophe. July has been their "boy" for fifteen years, and "his people" are his educated, good, white South African employers, as well as his own people, his black family and villagers deep in the country. In *July's People*, Gordimer portrays a future bloody South African revolution, which happily never took place. Instead there was the free election of 1994, the country narrowly drawing back from the brink of civil war.

Having stayed with Nadine Gordimer and her husband for several weeks at that time, I remember the joy, the laughter, and the hunger to see all and hear all of the miracles around us. "To have lived to see the end coming, and to have had some tiny part in it has been extraordinary and wonderful," she said in 1994. "It's like birth. As the baby's head is moulded by its passage down the birth canal so in South Africa your head, your mentality, your spirit, [are] forced into strange shapes by those extraordinary laws."

A Sport of Nature (1987) is Gordimer's most hazardous undertaking. Like *A Guest of Honour*, it is novel as history. Hillela runs off from an idyllic childhood to prove her sexuality and dive into the mess of the world's variety. Gordimer's empathy and affection are again with the blacks. It is a Cinderella tale summing up Africa's postcolonial history. Hillela is a despised daughter who enters palaces and presidencies through her political and sexual alliances. She marries an unscrupulous West African politician who becomes president of an African country and so attends the installation of the first South African black president (a thinly disguised Nelson Mandela). The finale is a vision of the future, but the focus is on Hillela as an honoured guest of a country where she was once a rebellious little white nobody. With Hillela, the intelligent, sensual heroine of a political picaresque, "Gordimer has met a fictional character she almost entirely loves," says her biographer Stephen Clingman. Gordimer took political and literary risks in this brutal fairy-tale of the dreadful year 1987, but she was right in predicting that liberation was only a few years away.

The Search for Identity

Nadine Gordimer's great themes are love and politics. Behind the most intimate relations, as well as the most public, there is the same search for an identity, a self-confirmation, and a wish to belong and exist. For Gordimer, the novel and the short story are instruments to penetrate a society that defends itself against scrutiny, hides in censorship and hypocrisy, refuses to recognise its history, and thus produces a grammar of lies where capitalism, liberalism, and Marxism mean the same thing:

an onslaught on the volk. She enters people's most intimate regions to show how private life is violated by informers and race registers. To write from within the personal sphere and make it public is the contrary to the police method of crashing into houses to confiscate letters and diaries, an act the teenaged Gordimer herself witnessed when the police raided a servant's room in her family's house.

Her characters live in the shadow of violence, threatened by unpredictable brutality. Races and classes, conventions and codes ferment in a decoction of final showdowns and a mysteriously glimmering hope of unexpected mergers and elective affinities outlined in the sands of the future. Through her language and fearless characterisation, Gordimer became a counterweight to the regime's propaganda. Unsentimental and diagnostic, she reports from the heart of darkness.

In a country that for so long feared new thoughts and orientations, Gordimer has scraped away the many layers of prejudice and egoism; she has dug out the fragile roots of a common fate and made us glimpse the brilliant colours of a world untainted by apartheid.

Characters

Her novel *The House Gun* (1998) is a morally complex, moving story from the liberated South Africa. A white murderer may now be defended by a black lawyer; in fact, it is this highly educated black man on whom two intelligent, well-read parents depend for the survival and sanity of their souls and for the redefinition of meaning in their lives. Their son has killed a man he loves, out of jealousy. Natalie, the mistress who is the impetus for the murder, is self-destructive and rebels against every form of personal dependence. It is a fable of violence and the search for new forms of freedom; it is also courtroom reportage. Had not the house gun been around, as it generally is in white families, no murder would have happened. This, in turn, evokes reflection on the fact of the general rise of violence in the world. The gun bought like any commodity in many countries — in the United States, Great Britain, France, or Japan — serves domestic violence and often falls into the hands of a child, with tragic consequences.

Gordimer's style and perspective, more complex in the latest novels, reflect on her words about the writer's dialectic: the tension between "excessive preoccupation and identification with the lives of others" and a "monstrous detachment." *The House Gun* has a definite voice of its own. Like *The Conservationist*, it stands stylistically apart from many of her other works. Who is the subject of this tragedy? Where are our edges? Where do the boundaries of self overlap, making each one responsible for the other's reality in a time of swift flux?

Gordimer is not afraid to present women of extraordinary intelligence and utmost delicacy of feeling as well as their vulgar counterparts. Take the human rights lawyer, Vera Stark, in *None to Accompany Me*. She is tormented by her husband Bennet's love, which remains the same without taking in what has happened around them. Vera wants someone who is committed to matters she thinks are important. She finds in work a defiant independence, which earlier she has experienced in her erotic life. Her foremost responsibility is with the liberation struggle and with her own sense of self.

Becoming free, Vera locks herself out from most of what other people want their freedom for. In the end, she persuades herself that only without Bennet can she become a genuine human being. She suffers pangs of conscience because, in his uninvolved innocence, he cannot understand her rebellion. It is the solitariness — none to accompany me — in the midst of her community activism and her work for victims of persecution that is the paradox of Vera's life. She is looking for a combat-free zone on a battlefield. In her comradeship with those who are risking their lives, Vera gets closer to her black colleagues than she does to her husband.

Gordimer's strength, here and elsewhere, is that she confronts bold and dangerous questions and gives them form without offering a ready answer. How can one keep one's hands clean while working against a dirty regime that does not shrink from using any means at its disposal? Does freedom consist in losing the past bit by bit? Why is there always someone who cannot afford to remember and others who are incapable of forgetting, however much they want to?

The Truth of Fiction

Nadine Gordimer has never written an autobiography or produced testimonies. She works in the imaginative dimension, always on an expedition into the mysteries of human experience. She does not appear "armed and dangerous," as her friend Ronnie Kasrils, one-time terrorist, later cabinet minister, was described by the police as late as 1992; but, in fact, she is, for hardly anyone has so vividly alerted the world to how apartheid undermined relations between people and made innocence criminal.

"Nothing I say in essays and articles will be as true as my fiction," she stated in an interview in *Transition* (no. 56, 1992). Because fiction is a disguise, it can "encompass all the things that go unsaid among other people and in yourself ... There is always, subconsciously, some kind of self-censorship in nonfiction." She added that, in a certain sense, a writer is selected by her subject, which is the consciousness of her own era.

Today, Nadine Gordimer lives and writes in a half-formed society of a kind almost never before seen on earth. Black and white have agreed to bring about a multiracial democracy by their faith as much as by their work. But the present stems from the past, and apartheid's contempt for human life now expresses itself in street killings, gang massacres, and armed robbery.

Gordimer's territory has always been the border between private emotions and external forces. There are no neutral zones where people can rest unobserved. In a land of lies, everyone lives a double life. Only love, the erotic dimension, stands for a sort of liberty, the glimpse of a more truthful existence. Outside the lovers' chamber, there is a society, greedy, immoral where empathy and responsibility for others, whatever skin colour, are rare. Thus, every meeting becomes instrumental or absurd. In many of her stories, Gordimer reminds us that the future of South Africa is not only a question of votes for all but one that requires immense effort to create a civil spirit, allowing people to look each other in the eye.

The responsibility of love and the loss of understanding, the loss of a grip on the world that comes with the end of love, are central themes in all of Gordimer's books. She is a moralist of a kind Alfred Nobel would have approved. She finds an uncommitted life not worth living. Her revolutionaries or human rights lawyers may have agonising personal problems, but they do not give up. In her later novels, there are people with energy and vision, as well as those who see nothing clearly — the former women, the latter often men. Gordimer seems to keep her characters at a distance in order to maintain a sense of the unknowable. Then one may discover, as André Brink says, "that one's very attempt at understanding or confronting the mystery opens up spaces of awareness one has not suspected before." Her true concerns reach beyond issues of the time to test the limits of human relationships and of language itself.

The Writer's Mission

Thanks to Nadine's and Reinhold's hospitality and our friendship of more than forty years, I have stayed in their house, built around 1910, longer than in anyone's house. It has hardly changed; I know every corner of it — her books, the paintings and the African handicraft she and Reinhold have collected over the years, the smells, the way to move in the kitchen and in the garden. The house is like the childhood retreat where I spent my summer holidays. A tree planted just before I visited the house for the first time is now huge. The police has never raided the house, although she has hidden some ANC fighters during a national alert for their seizure.

In her Nobel Lecture, Nadine Gordimer warned that an author risks both the state's condemnation as a traitor and the freedom movement's complaint that he or she fails to demonstrate blind loyalty. But, she explained, "the author serves humanity only as long as he utilises the word against his own loyalties too." Thus, a key paragraph in her book of Harvard lectures, *Writing and Being* (1994, p. 130) is this:

Only through a writer's explorations could I have begun to discover the human dynamism of the place I was born to and the time it was to be enacted. Only in the prescient dimension of the imagination could I bring together what had been deliberately broken and fragmented; fit together the shapes of living experience, my own and that of others, without which a whole consciousness is not attainable. I had to be part of the transformation of my place in order for it to know me.

Nadine Gordimer's work has grown into a profoundly psychological and social chronicle of half a century in South Africa. She is both its archivist and lighthouse keeper. Above her collected experience, the light sweeps, illuminating parts that would otherwise have lain in darkness, helping us navigate towards a South Africa that, far from being geographically cut off and politically ostracised, depicts a universal landscape.

❋

Derek WALCOTT

Photo courtesy of Anders Hallengren and the Nobel Foundation.

A Single, Homeless, Circling Satellite — Derek Walcott

✳

Jöran Mjöberg

Background and Youth

Derek Walcott was born in 1930 on St. Lucia, an island then belonging to the British Empire, but which became independent in 1979. St. Lucia has a hybrid British/French culture, having alternated as a colony of either England or France across the centuries. Walcott's ancestry is also mixed, with both his maternal and paternal grandmothers being black. His mother was a respected teacher at a Methodist infant school while his father died when Derek was only one year old.

His civil servant father had been an amateur painter, and the son has also devoted much of his grown-up life to painting, not to mention the many references to the great names in art all through his literary works. When growing up in Castries, the capital of St. Lucia, young Walcott attended St. Mary's College where his most important mentor was a painter, Harold Simmons. He soon took an interest in great European artists like Cézanne, Gauguin, and van Gogh.

While the town of Castries had an Europeanized culture, Afro-Caribbean folk customs and traditions dominated the countryside of

St. Lucia. Walcott published his first poem when he was just fourteen. At sixteen he wrote five plays and had his first collection of poetry published. By the age of twenty, Walcott was ready to found a theatre company on his own, the ST. LUCIA ARTS GUILD. In its inaugural year, this company produced his play *Henri-Christophe*, whose subject was taken from the colonial history of another Caribbean island, namely Haiti.

After graduating from St. Mary's College, Walcott continued his studies in another part of the Caribbean, on the island of Jamaica, where he attended the University College of the West Indies at Mona. Here he obtained his bachelor's degree in 1953. At the University College, he was both the editor of the student magazine and the president of MONA DRAMATIC SOCIETY.

The Search for an Identity

A central theme that runs throughout Walcott's works is his search for identity. From the beginning, he has intensely felt the antagonisms between the cultural heritage of the Old World and the traditions of the new one. In his critical work *Derek Walcott*, published in 1999, John Thieme describes the conflicts Walcott has experienced between the positions of European and African, Anglophone and Francophone, Standard English and Creole, and Methodist and Catholic. In the earlier collections of poetry, Thieme traces "a sense of lost perfection, cracked innocence and psychic fragmentation," which he considers to be a result of the racial divisions of the Caribbean society. In one volume after another, by means of a variety of important poems, Walcott tries to find expressions for the difficulties inherent in Caribbean identity. In "A Far Cry from Africa" (1962) he depicts his desperate dilemma in rather brutal formulations:

> *The gorilla wrestles with the superman.*
> *I who am poisoned with the blood of both,*
> *Where shall I turn, divided to the vein?*

Yet, in his fascinating essay of 1970, "What the Twilight Says," where he delivers a report about the origin of his interest in the theatre, he sounds more optimistic, hoping to be able to make creative use of his cultural schizophrenia.

In the poem "The Schooner Flight" (1979), Shabine, a Walcott persona, gives an often quoted definition of the identity of a person from a small country in the Caribbean:

> *I have Dutch, nigger, and English in me,*
> *and either I am nobody, or I'm a nation*

In reality, this meant:

> *I had no nation now but the imagination.*

In a somewhat later work, "North and South" (1981), this poem's persona gives another effort to express an identity, referring to himself as

> *a colonial upstart at the end of an empire,*
> *a single, homeless, circling satellite.*

At an early stage, Walcott was seized by an interest in the situation of St. Lucia. This grew into a promise to chronicle his island, a vow taken together with a painter friend. Walcott's early play, *Henri-Christophe*, was connected with this intense desire to depict and express the essence of his Caribbean surroundings.

In a later context, Walcott managed with deeper penetration than ever before to give form to a mature attitude to this theme, with a kind of acceptance of the trespasses of his ancestors through the centuries. Here follows the end and epitome of his extremely interesting essay "The Muse of History," published in 1976 and re-published in 1998 in the essays with the title "What the Twilight Says":

> *I accept this archipelago of the Americas, I say to the ancestor*
> *who sold me, and to the ancestor who bought me, I have no*
> *father, I want no such father, although I can understand*
> *you, black ghost, white ghost, when you both whisper "history,"*
> *for if I attempt to forgive you both I am falling into your*
> *idea of history which justifies and explains and expiates, and*
> *it is not mine to forgive, my memory cannot summon any*

filial love, since your features are anonymous and erased and
I have no wish and no power to pardon. You were when you
acted your roles, your given, historical roles of slave seller and
slave buyer, men acting as men, and also you, father in the
filth-ridden gut of the slave ship, to you they were also men,
your fellowman and tribesman not moved or hovering with
hesitation about your common race any longer than my other
bastard ancestor hovered with his whip, but to you, inwardly
forgiven grandfathers, I, like the more honest of my race, give
a strange thanks.

These are moving words for a person who feels himself exiled from the Eden of his grandfathers. We may be sure that this reconciliation has cost Walcott much but provided him with deep inner peace. But if we think of its universal consequences, this does not mean that there should exist any universal forgiveness for brutality. Thus, Walcott has no forgiveness when he asks in *Omeros* whether he might have broken his pen when he started writing poetry forty years earlier, if he had realized that

this century's pastorals were being written
by the chimneys of Dachau, of Auschwitz, of
Sachsenhausen

Explorers

Walcott is also fascinated by thoughts of the first men to discover and visit the world to which he belongs. These were explorers like Columbus, Walter Raleigh, and James Cook, as well as rebels like Toussaint and Henri-Christophe. To Walcott, Robinson Crusoe, more than anybody else, is a real archetype, and his long poem, "Crusoe's Island" (published in the 1965 volume *The Castaway*), contains in addition to a detailed geographic and psychological characterization, simple, lucid lines like the following ones:

Upon this rock the bearded hermit built
His Eden:

Goats, corn crop, fort, parasol, garden,
Bible for Sabbath, all the joys
But one
Which sent him howling for a human voice.
Exiled by a flaming sun
The rotting nut, bowled in the surf,
Became his own brain rotting from the guilt
Of heaven without his kind,
Crazed by such paradisal calm
The spinal shadow of a palm
Built keel and gunwale in his mind.

In the 1978 play *Pantomime*, Walcott used only two characters, Robinson and Friday, in an ironic, modernized variation of their personal relationship that takes place on the island of Tobago. In his important, autobiographical collection of poetry, *Another Life*, 1973, he also speaks about the task of those who first came over the seas to inhabit the American world:

We were blest with a virginal, unpainted world
with Adam's task of giving things their names

An important part of Walcott's poetry and drama has as a partly subconscious program, the "Caribbeanization" of earlier, European motives. Thus, when he studies and admires the plays of John Synge and his depiction of Aran fishermen, as well as the filmatic work of the Japanese director Akira Kurosawa, he works by creating St. Lucian counterparts, simple fishermen speaking their patois.

The Dramatic Work

Walcott's dramatic work is as important as his poetry. Today, he has written about twenty-five plays, although not all of them have been published. He has defined himself as "not only a playwright but a company," the reason being that he has worked as much as an instructor and as founder of theatre companies as a playwright. After starting "St. Lucia Arts Guild" in 1950, he opened "Little Carib Theatre

Workshop" in 1961. He had then hired a small troop of part-time actors, who could survive because they had other part-time occupations besides. They were nevertheless working under unsure economic conditions, with occasional contributions from the Rockefeller Foundation. In 1966, Walcott's company changed its name to "Trinidad Theatre Workshop". As its success gradually grew, the new company made guest performances abroad — in Jamaica, Guyana, Toronto (Canada), Boston, and New York (the USA). Walcott himself has worked at different places teaching, including Boston University as a professor of drama.

Among Walcott's earlier plays, *Ti-Jean and His Brothers* has a background in Caribbean folklore, while *Dream on Monkey Mountain*, his dramatic masterpiece, takes place on his own island of St. Lucia. The latter work's social inspiration derives from Jean-Paul Sartre's theories about the black Orpheus, as well as from Frantz Fanon, the French sociologist who impressed the peoples of the Western colonies so deeply with his work *Les Damnés de la Terre* (1956).

Dream on Monkey Mountain

The *Dream on Monkey Mountain* (1967) belongs to the twentieth-century genre called dream plays, connected with works by playwrights such as Strindberg as well as by Synge and Soyinka. The play's main character is Makak (French patois for "ape"), a black charcoal-burner who comes to town, gets drunk, and is taken into custody by Corporal Lestrade, a mulatto guard who is the maintainer of law and order during the later years of the colonial power. In a dream scene of a mock trial that was probably inspired by Kafka and Hesse, Lestrade accuses Makak of being intoxicated and damaging the premises of a local salesman. However, in another vivid dream sequence, Makak is crowned king in the romantic Africa of his roots, surrounded by his wives, his warriors, and the masks of pagan gods.

In a second mock trial, a number of great Western characters (e.g., Plato, Ptolemy, Dante, Cecil Rhodes, Florence Nightingale) are accused of neglecting other races and sentenced to death by the African tribes. Lestrade has now given up his confession to the Western world, shouldered his black inheritance, and sworn allegiance to Makak. The poor charcoal-burner is acquitted from the charges, and able to withdraw to his West Indian world with a deepened sense of identity.

The dream visions in this play seem to belong both to Makak and to the collective atmosphere of the plot. Ironic effects appear throughout the events. At the same time as Makak's romantic dream of Africa is presented, he cherishes a fantasy of a white protectress who takes care of him. But, as suggested by Lestrade, he gives up this dream, brutally beheading the woman with an African sword. This is a sacrifice that expresses a sound reaction against a fantasy life alienated from reality. Makak's character also bears symbolic similarities with Christ: in prison, he is followed by two robbers, and from Good Friday he is able to look forward to the moment of resurrection on Easter Sunday. The prison can be understood as a symbol both of life itself and of colonial rule. In a sophisticated way, this play expresses central components of Walcott's attitude to the political, racial, and psychological problems in his post-colonial world.

In *Dream on Monkey Mountain*, Walcott makes a great effort to interpret the nature of Caribbean identity. Colonialism has been important in damaging the human soul and humiliating the inhabitants of this part of the world. But there is no point trying to build castles in the air, as when Makak dreams of his African roots. At the end, in the epilogue, this simple-hearted visionary proletarian is acquitted, while Western civilization with its great characters is sentenced to death. Regardless of this, hate and revenge are negligible — in fact, negative — factors to the writer Walcott.

From Seville to Babylon

By *Ti-Jean and His Brothers* (1958), Walcott had more seriously started to embrace song and dance in the plot. He had been very successful with his first musicals, *The Joker of Seville* (1974) and *O Babylon* (1976). The former is a re-working of Tirso de Molina's play *El burlador de Sevilla*, and deals with the Don Juan character and its sexual and moral aspects, while at the same time taking up folk traditions and folk music (calypso) from Trinidad. *O Babylon* goes back to Walcott's experiences in Jamaica and deals with the opinions of a religious and political sect of this island, the Rastafarians, and their rejection of Western culture. In *Dream on Monkey Mountain*, the dances, the miming, and the masquerades take on an even wider role.

The "Homeric" Works

Omeros

From his early youth, Walcott had a great interest in both the sea and the Homeric world, calling the latter "an echo in the throat." Comparatively recently, he devoted two works to this subject: *Omeros* (1990) and *The Odyssey — A Stage Version* (1993). *Omeros* is a work divided into one hundred and ninety-two songs, written in a rhythmic blank verse with a richness of poetic metaphors and similes. In the French title, Walcott makes poetic pun in that *mer* evokes the sense of both "sea" and "mother," and "o" signifies the sound blown through a conch from the sea. This great work presents a reversible world, a colonial or post-colonial model corresponding to the original Homeric world. This is an epic poetic tale, with a multitude of different short stories, flashbacks, conversations, monologues, episodes, descriptions, and impressions, depicting in a minutely detailed way the Caribbean world and all its everyday life, its human beings, animals, nature, waters, and woods.

In *Omeros*, Homer himself appears in a row of different shapes. He is the blind Greek poet himself, the blind popular poet Seven Seas, the African griot or rhapsodist, the famous American painter Winslow Homer (with his paintings from the Atlantic Ocean), Virgil (the Roman counterpart to the Greek poet), and a blind barge-man who turns up on the stairs of the London church St. Martin-in-the-Fields with a manuscript refused by the editors. Even the characters correspond to the Homeric ones: Philoctete, the wounded archer; Major Plunkett, a contemporary Philoctete; Achilles, here the son of an African slave; Hector, a fisherman; Helen, intentionally made into a very common-place and approachable young Caribbean woman. Walcott's post-colonial world, a world where many slaves had classic Greek names, in many different ways corresponds to Rome and Greece. How could the poet, he says, while listening to the quarrel of two fishermen in his hometown, avoid thinking of quarreling Homeric characters?

Walcott's text is crowded with thoughts and reflections on history: "the farthest exclamations of history are written by a flag of smoke," exemplified by Troy, Carthage, Pompeii; "art is history's nostalgia," implying that literature carries the same guilt as history and history is midden built on midden.

Likewise, as a background to the life of people in our time, Walcott refers to violent events in history: the siege of Troy, the extermination of the Aruac people in the Caribbean by conquistadors, the eighteenth-century fights in the Caribbean between the English and French navies, as well as the prolonged catastrophe that extinguished most native Americans. Or the cruel attacks on African villages by slave traders, the perpetual tragedy of the captives who had to leave their homes, their families, their professions, and their tools, to try to create a new identity beyond the Atlantic: Ibos, Guineans, and many others.

The Odyssey

In a similar manner, the theatre production *The Odyssey* testifies to Walcott's deep interest, or rather involvement, in the Homeric world. There are, indeed, similarities between *Omeros* and *The Odyssey*, but there are also major differences. As a dramatic work, *The Odyssey* is divided into two acts, the first with fourteen scenes, the second with six. The speeches are short, usually only one line each, with the exception of the songs sung by the blind Billy Blue, who is a more modern version of Homer. Now and then, in a number of lines, the speeches have endings that form natural rhymes.

The characters are well-known from Homer and include Penelope, Odysseus' wife, who has to wait for his return from Troy for twenty years; Telemachus, his son; his old nurse on Ithaca, Eurycleia, who is the first to recognize him when he comes back at last from his many adventures; and Eumaeus, the shepherd. There are also the kings visited by Telemachus when he seeks his father: Nestor of Pylos and Menelaos of Sparta. We meet with the sailors of Odysseus' ship; King Alcinous and his daughter Nausicaa on the isle of the Phaeacians; Cyclops, the dangerous giant; Circe, the seductress; and in a short scene, corresponding to the sixth song of Homer's work, Odysseus' own mother Anticlea in the Underworld.

This does not mean, however, that all these characters are copies of those in the Greek Odyssey. Walcott is strikingly independent in forming different personalities. This work is not characterized by the same breadth and depth of the descriptions as in *Omeros*, but its dramatic verve, its liveliness, and its exquisite sense of humor distinguish it. We may accompany Odysseus from the victory at Troy, over his different stations on his way home, as well as we become more closely acquainted with Telemachus on his different expeditions and with Penelope in her difficult position in Ithaca. And the final scenes where Odysseus comes home and is at last recognized by Penelope and Telemachus do not lose any of the thrilling effects connected with the original Homeric situation. With its light, witty dialogue, it is in some ways more accessible than its poetic relative. Together, these two works provide some idea of Walcott's rich cultural and political outlook over the seas and continents of the human world.

※

Walcott writing.

Photo courtesy of Anders Hallengren and the Nobel Foundation.

Derek WALCOTT
Photo courtesy of Sigrid Nama and Derek Walcott.

The Antilles: Fragments of Epic Memory*

❋

Derek Walcott

Felicity is a village in Trinidad on the edge of the Caroni plain, the wide central plain that still grows sugar and to which indentured cane cutters were brought after emancipation, so the small population of Felicity is East Indian, and on the afternoon that I visited it with friends from America, all the faces along its road were Indian, which, as I hope to show, was a moving, beautiful thing, because this Saturday afternoon *Ramleela*, the epic dramatization of the Hindu epic the *Ramayana*, was going to be performed, and the costumed actors from the village were assembling on a field strung with different-coloured flags, like a new gas station, and beautiful Indian boys in red and black were aiming arrows haphazardly into the afternoon light. Low blue mountains on the horizon, bright grass, clouds that would gather colour before the light went. Felicity! What a gentle Anglo-Saxon name for an epical memory.

Under an open shed on the edge of the field, there were two huge armatures of bamboo that looked like immense cages. They were parts of the body of a god, his calves or thighs, which, fitted and reared,

---

* Nobel Lecture, December 7, 1992.

would make a gigantic effigy. This effigy would be burnt as a conclusion to the epic. The cane structures flashed a predictable parallel: Shelley's sonnet on the fallen statue of Ozymandias and his empire, that "colossal wreck" in its empty desert.

Drummers had lit a fire in the shed and they eased the skins of their tables nearer the flames to tighten them. The saffron flames, the bright grass, and the hand-woven armatures of the fragmented god who would be burnt were not in any desert where imperial power had finally toppled but were part of a ritual, evergreen season that, like the cane-burning harvest, is annually repeated, the point of such sacrifice being its repetition, the point of the destruction being renewal through fire.

Deities were entering the field. What we generally call "Indian music" was blaring from the open platformed shed from which the epic would be narrated. Costumed actors were arriving. Princes and gods, I supposed. What an unfortunate confession! "Gods, I suppose" is the shrug that embodies our African and Asian diasporas. I had often thought of but never seen *Ramleela*, and had never seen this theatre, an open field, with village children as warriors, princes, and gods. I had no idea what the epic story was, who its hero was, what enemies he fought, yet I had recently adapted the Odyssey for a theatre in England, presuming that the audience knew the trials of Odysseus, hero of another Asia Minor epic, while nobody in Trinidad knew any more than I did about Rama, Kali, Shiva, Vishnu, apart from the Indians, a phrase I use pervertedly because that is the kind of remark you can still hear in Trinidad: "apart from the Indians".

It was as if, on the edge of the Central Plain, there was another plateau, a raft on which the *Ramayana* would be poorly performed in this ocean of cane, but that was my writer's view of things, and it is wrong. I was seeing the *Ramleela* at Felicity as theatre when it was faith.

Multiply that moment of self-conviction when an actor, made-up and costumed, nods to his mirror before stopping on stage in the belief that he is a reality entering an illusion and you would have what I presumed was happening to the actors of this epic. But they were not actors. They had been chosen; or they themselves had chosen their roles in this sacred story that would go on for nine afternoons over a two-hour period till the sun set. They were not amateurs but believers. There was no theatrical term to define them. They did not have to psych themselves up to play their roles. Their acting would probably be as buoyant and as natural as those bamboo arrows crisscrossing the afternoon pasture. They believed in what they were playing, in the sacredness of the text, the validity of India, while I, out of the writer's habit, searched for some sense of elegy, of loss, even of degenerative mimicry in the happy faces of the boy-warriors or the heraldic profiles of the village princes. I was polluting the afternoon with doubt and with the patronage of admiration. I misread the event through a visual echo of History — the cane fields, indenture, the evocation of vanished armies, temples, and trumpeting elephants — when all around me there was quite the opposite: elation, delight in the boys' screams, in the sweets-stalls, in more and more costumed characters appearing; a delight of conviction, not loss. The name Felicity made sense.

Consider the scale of Asia reduced to these fragments: the small white exclamations of minarets or the stone balls of temples in the cane fields, and one can understand the self-mockery and embarrassment of those who see these rites as parodic, even degenerate. These purists look on such ceremonies as grammarians look at a dialect, as cities look on provinces and empires on their colonies. Memory that yearns to join the centre, a limb remembering the body from which it has been severed, like those bamboo thighs of the god. In other words, the way that the Caribbean is still looked at, illegitimate, rootless, mongrelized. "No people there", to quote Froude, "in the true sense of the word". No people. Fragments and echoes of real people, unoriginal and broken.

The performance was like a dialect, a branch of its original language, an abridgement of it, but not a distortion or even a reduction of its epic scale. Here in Trinidad I had discovered that one of the greatest epics of the world was seasonally performed, not with that desperate resignation of preserving a culture, but with an openness of belief that was as steady as the wind bending the cane lances of the Caroni plain. We had to leave before the play began to go through the creeks of the Caroni Swamp, to catch the scarlet ibises coming home at dusk. In a performance as natural as those of the actors of the *Ramleela*, we watched the flocks come in as bright as the scarlet of the boy archers, as the red flags, and cover an islet until it turned into a flowering tree, an anchored immortelle. The sigh of History meant nothing here. These two visions, the *Ramleela* and the arrowing flocks of scarlet ibises, blent into a single gasp of gratitude. Visual surprise is natural in the Caribbean; it comes with the landscape, and faced with its beauty, the sigh of History dissolves.

We make too much of that long groan which underlines the past. I felt privileged to discover the ibises as well as the scarlet archers of Felicity.

The sigh of History rises over ruins, not over landscapes, and in the Antilles there are few ruins to sigh over, apart from the ruins of sugar estates and abandoned forts. Looking around slowly, as a camera would, taking in the low blue hills over Port of Spain, the village road and houses, the warrior-archers, the god-actors and their handlers, and music already on the sound track, I wanted to make a film that would be a long-drawn sigh over Felicity. I was filtering the afternoon with evocations of a lost India, but why "evocations"? Why not "celebrations of a real presence"? Why should India be "lost" when none of these villagers ever really knew it, and why not "continuing", why not the perpetuation of joy in Felicity and in all the other nouns of the Central Plain: Couva, Chaguanas, Charley Village? Why was I not letting my pleasure open its windows wide? I was enticed like any Trinidadian to the ecstasies of their claim, because ecstasy was the pitch of the sinuous drumming in the loudspeakers. I was entitled to the feast of Husein, to the mirrors and crepe-paper temples of the Muslim epic, to the Chinese Dragon Dance, to the rites of that Sephardic Jewish synagogue that was once on Something Street. I am only one-eighth the writer I might have been had I contained all the fragmented languages of Trinidad.

Break a vase, and the love that reassembles the fragments is stronger than that love which took its symmetry for granted when it was whole. The glue that fits the pieces is the sealing of its original shape. It is such a love that reassembles our African and Asiatic fragments, the cracked heirlooms whose restoration shows its white scars. This gathering of broken pieces is the care and pain of the Antilles, and if the pieces are disparate, ill-fitting, they contain more pain than their original sculpture, those icons and sacred vessels taken for granted in their ancestral places. Antillean art is this restoration of our shattered histories, our shards of vocabulary, our archipelago becoming a synonym for pieces broken off from the original continent.

And this is the exact process of the making of poetry, or what should be called not its "making" but its remaking, the fragmented memory, the armature that frames the god, even the rite that surrenders it to a final pyre; the god assembled cane by cane, reed by weaving reed, line by plaited line, as the artisans of Felicity would erect his holy echo.

Poetry, which is perfection's sweat but which must seem as fresh as the raindrops on a statue's brow, combines the natural and the marmoreal; it conjugates both tenses simultaneously: the past and the present, if the past is the sculpture and the present the beads of dew or rain on the forehead of the past. There is the buried language and there is the individual vocabulary, and the process of poetry is one of excavation and of self-discovery. Tonally the individual voice is a dialect; it shapes its own accent, its own vocabulary and melody in defiance of an imperial concept of language, the language of Ozymandias, libraries and dictionaries, law courts and critics, and churches, universities, political dogma, the diction of institutions. Poetry is an island that breaks away from the main. The dialects of my archipelago seem as fresh to me as those raindrops on the statue's forehead, not the sweat made from the classic exertion of frowning marble, but the condensations of a refreshing element, rain and salt.

Deprived of their original language, the captured and indentured tribes create their own, accreting and secreting fragments of an old, an epic vocabulary, from Asia and from Africa, but to an ancestral, an ecstatic rhythm in the blood that cannot be subdued by slavery or indenture, while nouns are renamed and the given names of places accepted like Felicity village or Choiseul. The original language dissolves from the exhaustion of distance like fog trying to cross an ocean, but this process of renaming, of finding new metaphors, is the same process that the poet faces every morning of his working day, making his own tools like Crusoe, assembling nouns from necessity, from Felicity, even renaming himself. The stripped man is driven back to that self-astonishing, elemental force, his mind. That is the basis of the Antillean experience, this shipwreck of fragments, these echoes, these shards of a huge tribal vocabulary, these partially remembered customs, and they are not decayed but strong. They survived the Middle Passage and the *Fatel Rozack*, the ship that carried the first indentured Indians from the port of Madras to the cane fields of Felicity, that carried the chained Cromwellian convict and the Sephardic Jew, the Chinese grocer and the Lebanese merchant selling cloth samples on his bicycle.

And here they are, all in a single Caribbean city, Port of Spain, the sum of history, Trollope's "non-people". A downtown babel of shop signs and streets, mongrelized, polyglot, a ferment without a history, like heaven. Because that is what such a city is, in the New World, a writer's heaven.

A culture, we all know, is made by its cities.

Another first morning home, impatient for the sunrise — a broken sleep. Darkness at five, and the drapes not worth opening; then, in the sudden light, a cream-walled, brown-roofed police station bordered with short royal palms, in the colonial style, back of it frothing trees and taller palms, a pigeon fluttering into the cover of an cave, a rain-stained block of once-modern apartments, the morning side road into the station without traffic. All part of a surprising peace. This quiet happens with every visit to a city that has deepened itself in me. The flowers and the hills are easy, affection for them predictable; it is the architecture that, for the first morning, disorients. A return from American seductions used to make the traveller feel that something was missing, something was trying to complete itself, like the stained concrete apartments. Pan left along the window and the excrescences rear — a city trying to soar, trying to be brutal, like an American city in silhouette, stamped from the same mould as Columbus or Des Moines. An assertion of power, its decor bland, its air conditioning pitched to the point where its secretarial and executive staff sport competing cardigans; the colder the offices the more important, an imitation of another climate. A longing, even an envy of feeling cold.

In serious cities, in grey, militant winter with its short afternoons, the days seem to pass by in buttoned overcoats, every building appears as a barracks with lights on in its windows, and when snow comes, one has the illusion of living in a Russian novel, in the nineteenth century, because of the literature of winter. So visitors to the Caribbean must feel that they are inhabiting a succession of postcards. Both climates are shaped by what we have read of them. For tourists, the sunshine cannot be serious. Winter adds depth and darkness to life as well as to literature, and in the unending summer of the tropics not even poverty or poetry (in the Antilles poverty is poetry with a V, une vie, a condition of life as well as of imagination) seems capable of being profound because the nature around it is so exultant, so resolutely ecstatic, like its music. A culture based on joy is bound to be shallow. Sadly, to sell itself, the Caribbean encourages the delights of mindlessness, of brilliant vacuity, as a place to flee not only winter but that seriousness that comes only out of culture with four seasons. So how can there be a people there, in the true sense of the word?

They know nothing about seasons in which leaves let go of the year, in which spires fade in blizzards and streets whiten, of the erasures of whole cities by fog, of reflection in fireplaces; instead, they inhabit a geography whose rhythm, like their music, is limited to two stresses: hot and wet, sun and rain, light and shadow, day and night, the limitations of an incomplete metre, and are therefore a people incapable of the subtleties of contradiction, of imaginative complexity. So be it. We cannot change contempt.

Ours are not cities in the accepted sense, but no one wants them to be. They dictate their own proportions, their own definitions in particular places and in a prose equal to that of their detractors, so that now it is not just St. James but the streets and yards that Naipaul commemorates, its lanes as short and brilliant as his sentences; not just the noise and jostle of Tunapuna but the origins of C. L. R. James's *Beyond a Boundary*, not just Felicity village on the Caroni plain, but Selvon Country, and that is the way it goes up the islands now: the old Dominica of Jean Rhys still very much the way she wrote of it; and the Martinique of the early Césaire; Perse's Guadeloupe, even without the pith helmets and the mules; and what delight and privilege there was in watching a literature — one literature in several imperial languages, French, English, Spanish — bud and open island after island in the early morning of a culture, not timid, not derivative, any more than the hard white petals of the frangipani are derivative and timid. This is not a belligerent boast but a simple celebration of inevitability: that this flowering had to come.

On a heat-stoned afternoon in Port of Spain, some alley white with glare, with love vine spilling over a fence, palms and a hazed mountain appear around a corner to the evocation of Vaughn or Herbert's "that shady city of palm-trees", or to the memory of a Hammond organ from a wooden chapel in Castries, where the congregation sang "Jerusalem, the Golden". It is hard for me to see such emptiness as desolation. It is that patience that is the width of Antillean life, and the secret is not to ask the wrong thing of it, not to demand of it an ambition it has no interest in. The traveller reads this as lethargy, as torpor.

Here there are not enough books, one says, no theatres, no museums, simply not enough to do. Yet, deprived of books, a man must fall back on thought, and out of thought, if he can learn to order it, will come the urge to record, and in extremity, if he has no means of recording, recitation, the ordering of memory which leads to metre, to commemoration. There can be virtues in deprivation, and certainly one virtue is salvation from a cascade of high mediocrity, since books are now not so much created as remade. Cities create a culture, and all we have are these magnified market towns, so what are the proportions of the ideal Caribbean city? A surrounding, accessible countryside with leafy suburbs, and if the city is lucky, behind it, spacious plains. Behind it, fine mountains; before it, an indigo sea. Spires would pin its centre and around them would be leafy, shadowy parks. Pigeons would cross its sky in alphabetic patterns, carrying with them memories of a belief in augury, and at the heart of the city there would be horses, yes, horses, those animals last seen at the end of the nineteenth century drawing broughams and carriages with top-hatted citizens, horses that live in the present tense without elegiac echoes from their hooves, emerging from paddocks at the Queen's Park Savannah at sunrise, when mist is unthreading from the cool mountains above the roofs, and at the centre of the city seasonally there would be races, so that citizens could roar at the speed and grace of these nineteenth-century animals. Its docks, not obscured by smoke or deafened by too. much machinery, and above all, it would be so racially various that the cultures of the world — the

Asiatic, the Mediterranean, the European, the African — would be represented in it, its humane variety more exciting than Joyce's Dublin. Its citizens would intermarry as they chose, from instinct, not tradition, until their children find it increasingly futile to trace their genealogy. It would not have too many avenues difficult or dangerous for pedestrians, its mercantile area would be a cacophony of accents, fragments of the old language that would be silenced immediately at five o'clock, its docks resolutely vacant on Sundays.

This is Port of Spain to me, a city ideal in its commercial and human proportions, where a citizen is a walker and not a pedestrian, and this is how Athens may have been before it became a cultural echo.

The finest silhouettes of Port of Spain are idealizations of the craftsman's handiwork, not of concrete and glass, but of baroque woodwork, each fantasy looking more like an involved drawing of itself than the actual building. Behind the city is the Caroni plain, with its villages, Indian prayer flags, and fruit vendors' stalls along the highway over which ibises come like floating flags. Photogenic poverty! Postcard sadnesses! I am not re-creating Eden; I mean, by "the Antilles", the reality of light, of work, of survival. I mean a house on the side of a country road, I mean the Caribbean Sea, whose smell is the smell of refreshing possibility as well as survival. Survival is the triumph of stubborness, and spiritual stubborness, a sublime stupidity, is what makes the occupation of poetry endure, when there are so many things that should make it futile. Those things added together can go under one collective noun: "the world".

This is the visible poetry of the Antilles, then. Survival.

If you wish to understand that consoling pity with which the islands were regarded, look at the tinted engravings of Antillean forests, with their proper palm trees, ferns, and waterfalls. They have a civilizing decency, like Botanical Gardens, as if the sky were a glass ceiling under which a colonized vegetation is arranged for quiet walks and carriage rides. Those views are incised with a pathos that guides the engraver's tool and the topographer's pencil, and it is this pathos which, tenderly ironic, gave villages names like Felicity. A century looked at a landscape furious with vegetation in the wrong light and with the wrong eye. It is such pictures that are saddening rather than the tropics itself. These delicate engravings of sugar mills and harbours, of native women in costume, are seen as a part of History, that History which looked over the shoulder of the engraver and, later, the photographer. History can alter the eye and the moving hand to conform a view of itself; it can rename places for the nostalgia in an echo; it can temper the glare of tropical light to elegiac monotony in prose, the tone of judgement in Conrad, in the travel journals of Trollope.

These travellers carried with them the infection of their own malaise, and their prose reduced even the landscape to melancholia and self-contempt. Every endeavor is belittled as imitation, from architecture to music. There was this conviction in Froude that since History is based on achievement, and since the history of the Antilles was so genetically corrupt, so depressing in its cycles of massacres, slavery, and indenture, a culture was inconceivable and nothing could ever be created in those ramshackle ports, those monotonously feudal sugar estates. Not only the light and salt of Antillean mountains defied this, but the demotic vigour and variety of their inhabitants. Stand close to a waterfall and you will stop hearing its roar. To be still in the nineteenth century, like horses, as Brodsky has written, may not be such a bad deal, and much of our life in the Antilles still seems to be in the rhythm of the last century, like the West Indian novel.

By writers even as refreshing as Graham Greene, the Caribbean is looked at with elegiac pathos, a prolonged sadness to which Lévi-Strauss has supplied an epigraph: *Tristes Tropiques*. Their tristesse derives from an attitude to the Caribbean dusk, to rain, to uncontrollable vegetation, to the provincial ambition of Caribbean cities where brutal replicas of modern architecture dwarf the small houses and streets. The mood is understandable, the melancholy as contagious as the fever of a sunset, like the gold fronds of diseased coconut palms, but there is something alien and ultimately wrong in the way such a sadness, even a morbidity, is described by English, French, or some of our exiled writers. It relates to a misunderstanding of the light and the people on whom the light falls.

These writers describe the ambitions of our unfinished cities, their unrealized, homiletic conclusion, but the Caribbean city may conclude just at that point where it is satisfied with its own scale, just as

Caribbean culture is not evolving but already shaped. Its proportions are not to be measured by the traveller or the exile, but by its own citizenry and architecture. To be told you are not yet a city or a culture requires this response. I am not your city or your culture. There might be less of *Tristes Tropiques* after that.

Here, on the raft of this dais, there is the sound of the applauding surf: our landscape, our history recognized, "at last". *At Last* is one of the first Caribbean books. It was written by the Victorian traveller Charles Kingsley. It is one of the early books to admit the Antillean landscape and its figures into English literature. I have never read it but gather that its tone is benign. The Antillean archipelago was there to be written about, not to write itself, by Trollope, by Patrick Leigh-Fermor, in the very tone in which I almost wrote about the village spectacle at Felicity, as a compassionate and beguiled outsider, distancing myself from Felicity village even while I was enjoying it. What is hidden cannot be loved. The traveller cannot love, since love is stasis and travel is motion. If he returns to what he loved in a landscape and stays there, he is no longer a traveller but in stasis and concentration, the lover of that particular part of earth, a native. So many people say they "love the Caribbean", meaning that someday they plan to return for a visit but could never live there, the usual benign insult of the traveller, the tourist. These travellers, at their kindest, were devoted to the same patronage, the islands passing in profile, their vegetal luxury, their backwardness and poverty. Victorian prose dignified them. They passed by in beautiful profiles and were forgotten, like a vacation.

Alexis Léger, whose writer's name is Saint-John Perse, was the first Antillean to win this prize for poetry. He was born in Guadeloupe and wrote in French, but before him, there was nothing as fresh and clear in feeling as those poems of his childhood, that of a privileged white child on an Antillean plantation, *Pour fêter une enfance*, *Éloges*, and later *Images à Crusoé*. At last, the first breeze on the page, salt-edged and self-renewing as the trade winds, the sound of pages and palm trees turning as "the odour of coffee ascents the stairs".

Caribbean genius is condemned to contradict itself. To celebrate Perse, we might be told, is to celebrate the old plantation system, to celebrate the beque or plantation rider, verandahs and mulatto servants, a white French language in a white pith helmet, to celebrate a rhetoric of patronage and hauteur; and even if Perse denied his origins, great writers often have this folly of trying to smother their source, we cannot deny him any more than we can the African Aimé Césaire. This is not accommodation, this is the ironic republic that is poetry, since, when I see cabbage palms moving their fronds at sunrise, I think they are reciting Perse.

The fragrant and privileged poetry that Perse composed to celebrate his white childhood and the recorded Indian music behind the brown young archers of Felicity, with the same cabbage palms against the same Antillean sky, pierce me equally. I feel the same poignancy of pride in the poems as in the faces. Why, given the history of the Antilles, should this be remarkable? The history of the world, by which of course we mean Europe, is a record of intertribal lacerations, of ethnic cleansings. At last, islands not written about but writing themselves! The palms and the Muslim minarets are Antillean exclamations. At last! the royal palms of Guadeloupe recite *Éloges* by heart.

Later, in *Anabase*, Perse assembled fragments of an imaginary epic, with the clicking teeth of frontier gates, barren wadis with the froth of poisonous lakes, horsemen burnoosed in sandstorms, the opposite of cool Caribbean mornings, yet not necessarily a contrast any more than some young brown archer at Felicity, hearing the sacred text blared across the flagged field, with its battles and elephants and monkey-gods, in a contrast to the white child in Guadeloupe assembling fragments of his own epic from the lances of the cane fields, the estate carts and oxens, and the calligraphy of bamboo leaves from the ancient languages, Hindi, Chinese, and Arabic, on the Antillean sky. From the *Ramayana* to Anabasis, from Guadeloupe to Trinidad, all that archaeology of fragments lying around, from the broken African kingdoms, from the crevasses of Canton, from Syria and Lebanon, vibrating not under the earth but in our raucous, demotic streets.

A boy with weak eyes skims a flat stone across the flat water of an Aegean inlet, and that ordinary action with the scything elbow contains the skipping lines of the *Iliad* and the *Odyssey*, and another child aims a bamboo arrow at a village festival, another hears the rustling march of cabbage palms in a Caribbean sunrise, and from that sound, with its fragments of tribal myth, the compact expedition of Perse's epic is launched, centuries and archipelagoes apart. For every poet it is always morning in the world. History a forgotten, insomniac night; History and elemental awe are always our early beginning, because the fate of poetry is to fall in love with the world, in spite of History.

There is a force of exultation, a celebration of luck, when a writer finds himself a witness to the early morning of a culture that is defining itself, branch by branch, leaf by leaf, in that self-defining dawn, which is why, especially at the edge of the sea, it is good to make a ritual of the sunrise. Then the noun, the "Antilles" ripples like brightening water, and the sounds of leaves, palm fronds, and birds are the sounds of a fresh dialect, the native tongue. The personal vocabulary, the individual melody whose metre is one's biography, joins in that sound, with any luck, and the body moves like a walking, a waking island.

This is the benediction that is celebrated, a fresh language and a fresh people, and this is the frightening duty owed.

I stand here in their name, if not their image — but also in the name of the dialect they exchange like the leaves of the trees whose names are suppler, greener, more morning-stirred than English — *laurier canelles, bois-flot, bois-canot* — or the valleys the trees mention — *Fond St. Jacques, Matoonya, Forestier, Roseau, Mahaut* — or the empty beaches — *L'Anse Ivrogne, Case en Bas, Paradis* — all songs and histories in themselves, pronounced not in French — but in patois.

One rose hearing two languages, one of the trees, one of school children reciting in English:

> *I am monarch of all I survey,*
> *My right there is none to dispute;*
> *From the centre all round to the sea*
> *I am lord of the fowl and the brute.*
> *Oh, solitude! where are the charms*
> *That sages have seen in thy face?*
> *Better dwell in the midst of alarms,*
> *Than reign in this horrible place...*

While in the country to the same metre, but to organic instruments, handmade violin, chac-chac, and goatskin drum, a girl named Sensenne singing:

> *Si mwen di 'ous ça fait mwen la peine*
> *'Ous kai dire ça vrai.*
> (If I told you that caused me pain
> You'll say, "It's true".)
> *Si mwen di 'ous ça pentetrait mwen*
> *'Ous peut dire ça vrai*
> (If I told you you pierced my heart
> You'd say, "It's true".)
> *Ces mamailles actuellement*
> *Pas ka faire l 'amour z'autres pour un rien.*
> (Children nowadays
> Don't make love for nothing.)

It is not that History is obliterated by this sunrise. It is there in Antillean geography, in the vegetation itself. The sea sighs with the drowned from the Middle Passage, the butchery of its aborigines, Carib and Aruac and Taino, bleeds in the scarlet of the immortelle, and even the actions of surf on sand cannot erase the African memory, or the lances of cane as a green prison where indentured Asians, the ancestors of Felicity, are still serving time.

That is what I have read around me from boyhood, from the beginnings of poetry, the grace of effort. In the hard mahogany of woodcutters: faces, resinous men, charcoal burners; in a man with a cutlass cradled across his forearm, who stands on the verge with the usual anonymous khaki dog; in the extra clothes he put on this morning, when it was cold when he rose in the thinning dark to go and make his garden in the heights — the heights, the garden, being miles away from his house, but that is where he has his land — not to mention the fishermen, the footmen on trucks, groaning up mornes, all fragments of Africa originally but shaped and hardened and rooted now in the island's life, illiterate in the way leaves are illiterate; they do not read, they are there to be read, and if they are properly read, they create their own literature.

But in our tourist brochures the Caribbean is a blue pool into which the republic dangles the extended foot of Florida as inflated rubber islands bob and drinks with umbrellas float towards her on a raft. This is how the islands from the shame of necessity sell themselves; this is the seasonal erosion of their identity, that high-pitched repetition of the same images of service that cannot distinguish one island from the other, with a future of polluted marinas, land deals negotiated by ministers, and all of this conducted to the music of Happy Hour and the rictus of a smile. What is the earthly paradise for our visitors? Two weeks without rain and a mahogany tan, and, at sunset, local troubadours in straw hats and floral shirts beating "Yellow Bird" and "Banana Boat Song" to death. There is a territory wider than this — wider than the limits made by the map of an island — which is the illimitable sea and what it remembers.

All of the Antilles, every island, is an effort of memory; every mind, every racial biography culminating in amnesia and fog. Pieces of sunlight through the fog and sudden rainbows, arcs-en-ciel. That is the effort, the labour of the Antillean imagination, rebuilding its gods from bamboo frames, phrase by phrase.

Decimation from the Aruac downwards is the blasted root of Antillean history, and the benign blight that is tourism can infect all of those island nations, not gradually, but with imperceptible speed, until each rock is whitened by the guano of white-winged hotels, the arc and descent of progress.

Before it is all gone, before only a few valleys are left, pockets of an older life, before development turns every artist into an anthropologist or folklorist, there are still cherishable places, little valleys that do not echo with ideas, a simplicity of rebeginnings, not yet corrupted by the dangers of change. Not nostalgic sites but occluded sanctities as common and simple as their sunlight. Places as threatened by this prose as a headland is by the bulldozer or a sea almond grove by the surveyor's string, or from blight, the mountain laurel.

One last epiphany: A basic stone church in a thick valley outside
Soufrière, the hills almost shoving the houses around into a brown river,
a sunlight that looks oily on the leaves, a backward place, unimportant,
and one now being corrupted into significance by this prose. The idea is
not to hallow or invest the place with anything, not even memory.
African children in Sunday frocks come down the ordinary concrete
steps into the church, banana leaves hang and glisten, a truck is parked
in a yard, and old women totter towards the entrance. Here is where a
real fresco should be painted, one without importance, but one with
real faith, mapless, Historyless.

How quickly it could all disappear! And how it is beginning to drive
us further into where we hope are impenetrable places, green secrets at
the end of bad roads, headlands where the next view is not of a hotel
but of some long beach without a figure and the hanging question of
some fisherman's smoke at its far end. The Caribbean is not an idyll, not
to its natives. They draw their working strength from it organically, like
trees, like the sea almond or the spice laurel of the heights. Its peasantry
and its fishermen are not there to be loved or even photographed; they
are trees who sweat, and whose bark is filmed with salt, but every day
on some island, rootless trees in suits are signing favourable tax breaks
with entrepreneurs, poisoning the sea almond and the spice laurel of the
mountains to their roots. A morning could come in which governments
might ask what happened not merely to the forests and the bays but to a
whole people.

They are here again, they recur, the faces, corruptible angels,
smooth black skins and white eyes huge with an alarming joy, like those
of the Asian children of Felicity at *Ramleela*; two different religions,
two different continents, both filling the heart with the pain that is joy.

But what is joy without fear? The fear of selfishness that, here on
this podium with the world paying attention not to them but to me, I
should like to keep these simple joys inviolate, not because they are
innocent, but because they are true. They are as true as when, in the
grace of this gift, Perse heard the fragments of his own epic of Asia
Minor in the rustling of cabbage palms, that inner Asia of the soul
through which imagination wanders, if there is such a thing as
imagination as opposed to the collective memory of our entire race, as

true as the delight of that warrior-child who flew a bamboo arrow over the flags in the field at Felicity; and now as grateful a joy and a blessed fear as when a boy opened an exercise book and, within the discipline of its margins, framed stanzas that might contain the light of the hills on an island blest by obscurity, cherishing our insignificance.

❄

Naguib MAHFOUZ
Photo courtesy of Mohamed Hegazy.

Naguib Mahfouz – The Son of Two Civilizations

❋

Anders Hallengren

I am the son of two civilizations that at a certain age in history have formed a happy marriage. The first of these, seven thousand years old, is the Pharaonic civilization; the second, one thousand four hundred years old, is the Islamic civilization.

Naguib Mahfouz, Nobel Lecture

The Arabic Renaissance and the Rise of the Egyptian Novel

Arabic literature can be traced back almost two thousand years. Poetry has always been its most prominent genre, but there is also an ancient tradition of narrative that expresses itself in a wealth of different oral forms. In Egypt, the collection of stories called *The Arabian Nights*, a series of tales of Indian, Iranian, and Iraqi origin, was brought to its final and most developed form. This coincided with an ancient Egyptian tradition of storytelling, which has remained vivid and alive to this day, the public storyteller having been a cultural institution for ages.

The birth of the Egyptian novel, however, could not take place until the modern era, when five preconditions had been fulfilled: 1) the influence of European literature, where the novel developed into a major genre in the eighteenth and nineteenth centuries; 2) the establishment of Egyptian printing works and pressrooms in the nineteenth century along with the rise of newspaper production; 3) public education and the spread of literacy; 4) a gradual liberation from oppression by foreign powers, starting with the reign of Muhammad Ali in the aftermath of the Napoleonic occupation in the early 1800s; and 5) the emergence of an intellectual class with broad international learning.

Thus an Arabic Renaissance finally arose; its Janus face turned as much to the past as to the future. The concept of *Nahdah* in Arabic literary criticism and historiography, meaning a "rising up" or revitalisation, refers in part to a period of neo-classicism, an awakening of old literary traditions following a time of decline or stagnation since the eleventh century. The term also refers to creativity, new syntheses, modernisation, dynamic experiments, and progress (— as described by, for instance, J. Brugman in *An Introduction to the History of Modern Arabic Literature in Egypt*, Leiden: E. J. Brill, 1984, Ch. "The Nahdah"). The Egyptian novel matured in great works by twentieth century writers such as Muhammad Husayn Haykal (1888–1956), Taha Husayn (1889–1973), Ibrahim al-Mazini (1890–1949), Mahmud Tahir Lashin (1894–1954), and Tawfiq al-Hakim (1898–1987).

Muhammad Husayn Haykal's novel *Zaynab*, published in 1912, is often regarded as the first true Arabic novel, but there were many forerunners. The most important successor, however, was born about the time of the completion of Husayn Haykal's work. The development of the modern Egyptian novel is reflected by — and reaches a peak in — the half century of work by Naguib Mahfouz (1911–), Nobel Laureate in 1988, the first writer in Arabic to receive the Nobel Prize in Literature.

Mahfouz's Perennial Quest

Many Egyptians know the stories by Mahfouz from the cinema. This fact brings the central facets of the continuation of the so-called

Renaissance into immediate focus. The introduction of new genres and new media provided new means of expression, reflection, and creativity, and, in the opposite direction, influenced the creators of literary arts and genres as well as society as a whole. Taking an active part in the Egyptian film industry, both as an official and as a producer of manuscripts, Mahfouz is also part and parcel of a modernisation process that comprises the leading film-producing country in the Arab world. The Egyptian film industry is, next to that of India and the USA, the largest in the world.

The development of his writings is also connected with the constantly growing importance of news media, magazines, daily papers since Mahfouz himself is, for example, a contributor to and reader of the Cairo newspaper *Al-Ahram* (The Pyramids), founded in 1875.

In *Layali alf layla* (1982), translated as *Arabian Nights and Days*, Mahfouz, adopting the story telling fashion of *The Arabian Nights*, writes riddles which can be read against the background of contemporary politics and society, in particular perhaps the open-door policy introduced by President Sadat. Using the traditional types — ruler, ruled, the mystic guide, the secret connection — his tales are open to various interpretations. They are principally stories of good and evil, right and wrong, truth and betrayal, which make them of universal concern. They tell us of dreams and teach manners, as tradition has always done. Ultimately this collection of stories has, like classical fables, a moral status. However, the author is never a fabulist in any other sense, but looks upon himself as having a mission and a deep responsibility as a writer. In a way, Naguib Mahfouz seems to have taken the place of honour among ancient Egyptian scribes. As in the *Teaching of Cheti*, probably from the twelfth dynasty, in which books are praised as being "of superior value", as essential "as water", the writer has a moral calling. Even the Arabic word for literature, *adab*, originally refers to a high level of culture, i.e., good behaviour and exalted manners. Novels (*rumaniyat*) were originally also accepted and developed for the teaching of *adab*. The best-known stories by Mahfouz depict the lives and manners of lower middle-class families in Cairo and their environment. In particular, he tells us about the alleys of the El Gamaleya area (al-Jamaliyya), the neighbourhood in which he spent his childhood and youth.

To see his works as mainly political fables or allegories is fallacious. It is a most misleading simplification, since there are many levels of interpretation and reception. His novels and short stories are works of art. They picture Egyptian milieus from the most ancient of times to contemporary everyday life, deal with questions of broad human concern, raise philosophical and existential questions. The author is always guided by a belief in Egyptian continuity and greatness, from time to time shaken to its foundations by tumultuous history, the corruption of thought and disaster. In his novels there is a staunch belief in moral right and a constant seeking for Egyptian identity behind the weft of illusion and reality. A dweller in truth, unable to define it, Mahfouz is — like the investigator, Meriamun, in his novel on the enigmatic Akhenaten — perpetually pursuing his own self.

In the narrative stream of his short stories and his novels, the reader encounters a great variety of characters, people described as soon as they appear before us (or the other way round). They leave lasting impressions but also hold back something essential that does not come within our grasp. They turn up and disappear, leaving traces and clues, but remain enigmatic, ambiguous. They are figures in a greater story, or pieces in a puzzle, that is the œuvre of Mahfouz. Their lives are texts, continually being written and rewritten, as is Egyptian history. Their appearance changes as the context alters with time and setting. Likewise their meaning and purport depend upon viewpoint and perspective, and there are many layers of interpretation, from the gross to the subtle and inexpressible. A correct hermeneutics or a right understanding is as evasive as *al-sarab*, the mirage of *al-sahra*, the nothingness of a vast expanse of desert. So do human illusions appear before us, materialise, and fade away, leaving voids pregnant with meaning.

The Pertinence of the Past

Naguib Mahfouz started his career as a writer by exploring ancient Egyptian history. He did not do so to understand the contemporary scene, still less was it to criticise it in a covert fashion. His aim was to seek the identity of his own country in the space-time of his existence and the sphere of his Self. He also obviously sought for a reliable anchorage in the distant past during years of war, upheaval, and calamity. Being an Arabic author, he transcends the limits of Arabic and Moslem tradition, to which he belongs, tracing his heritage and seeking his identity as an Egyptian.

His first published book was a translation of James Baikie's concise history of Ancient Egypt (*Misr Al-Qadimah*, 1932), published by the periodical *Al-Majalah Al-Jadidah* (The New Magazine). Between 1939–1945, that is, during the years of the Second World War, he published three novels about ancient Egypt. His first published novel *'Abath al-Aqdar* (1939), "Ironies of Fate", dealing with "the malediction of Ra", was followed by two other historical novels about ancient Egypt. It was the beginning of a gigantic plan for forty projected novels. The author's aim was to employ the novel form to relate the history of Egypt from the earliest times up to his own day. However, writing about the past, he found himself writing about the present. Then he turned to the life of modern Egyptians, first of all to those in his own Cairo quarters. Later on he was to find that when writing about the present he was also writing about the past. It is no mere coincidence that he returns to the perspective of ancient Egypt in novels written in the 1980s, forty years later. In a way, the history of Egypt from past to present is the framework of all his writing.

The *'Abath al-Aqdar* (1939), is based on an ancient Egyptian legend. Originally the novel was entitled *Hikmat Khufu*, "The Wisdom of Cheops". This pharaoh, who lived about 2680 BC, during the fourth dynasty of the Old Kingdom, has been told by a soothsayer that after his death, his son will not inherit the kingdom. Instead, it will fall in the hands of Dedef, the son of the high priest of the temple of Ra. Cheops consequently takes every measure to change the future. In this drama of confusion, as intricate as the fate of King Oedipus and the legend of Moses, all attempts to interfere with destiny prove to be in vain.

It is notable that when the author later on turns to social and contemporary issues, something of this early belief in fate, destiny, dispensation or providence remains. People in his novels are often, like reeds in the wind, almost powerless in the face of circumstance and chance.

The second historical novel, *Radubis* (1943), is a story about a courtesan in ancient Egypt, and is named after her. She is the lover and mistress of the Pharaoh Mernere II, and the novel is about the romance between them. Not much is known about kings called Merenre or Mernere in the most ancient list of kings; we hardly know more about this king than that he ruled for a year at the end of the sixth dynasty, and that there was a queen Nitocris at the time, famous for her beauty.

These facts make the subject different from the first novel, giving imagination an even freer rein. The love between the pharaoh and the courtesan Radubis is a pure love of devotion. The gods sanction it, and at last he gives up all resistance, driven by forces beyond his control. Finally, he uses the assets of the country to celebrate the wonderful woman who captivated the heart of the pharaoh by means of divine forces.

In his book about the Hyksos wars, the author turns his attention to the decline and fall of the Middle Kingdom during its last period (1785–1575 BC). *Kifah Tiba* (1944), "The Struggle of Thebes", describes the wars against a foreign power of Asiatic origin which ruled over Egypt for a long time. It is a story of the end of an era. But when the Hyksos were finally expelled, Egypt became independent and began building a new and prosperous empire, The New Kingdom. The historical framework is itself a story of archetypal dimensions. Mahfouz is writing about a renaissance, the rise of Egyptian nationalism, the fight for independence and regained self-confidence reflected in his own times, as many times before.

Power and Social Issues

The series of historical books was followed by records of modern Cairo, the *Khan al-Khalili* (1946), the author turning abruptly but quite naturally to his own era. The novel describes the life and final tragedy of a family which, during the war, was forced to move to a less fashionable part of the city. The narrator-seeker walks freely and unhindered, as it were, in the same places in different times, sensing continuity, in pursuit of the red thread in the story that is made up by history and in which he is himself a part.

In many of the novels by Mahfouz, women play a central part. Western critics have sometimes remarked that novels are built up round women who can match the men in the narrative, and there are many of them in the Egyptian novels. In Mahfouz's works, prostitutes and other fallen females are often the strongest and the wisest characters. Mothers and other women constitute a secret net of devotion, passion, and care that holds the chaotic world of men together. The author's sympathy is with the oppressed and the miserable, with the weak and the loving. Women are often viewed as victims, as the victims of cruel circumstances, even when they are blamed and mocked. And that is true of men, too.

After the *Khan al-Khalili*, named after a working class area, he wrote *Zuqaq al-midaqq* (1947), "Midaq Alley", named in a similar fashion after a street in Cairo. The main character is a lower-class woman during the Second World War. A hairdresser is madly in love with her, but the open-handed British soldiers in the city also tempt her. Her Egyptian lover is finally crushed by unhappy love and jealousy, and she ends up as a lady of the night.

A special portrait of motherhood was presented in the next novel belonging to this socially concerned and realistic series of contemporary novels leading up to the Cairo Trilogy. *Al-Sarab* (1948), "The Mirage", tells about the crucial point where maternal instincts, motherly love and mother-fixation intersect. It is a psychological novel but, as many times before and afterwards, issues of fate, destiny and free will are raised but not answered. There is shame, but there is no guilt.

A mother who cannot let her son go after her husband/his father has left her, brings up the protagonist. He is the apple of her eye, and there is no lack of love during the years when he is growing up. He is unhappy and miserable at school and at the university, however, and he gets an insignificant government post. He marries, but cannot establish a good relation with his wife, who is unfaithful, she has a love affair, and dies after an abortion. When he finally dares to revolt against his mother and reproaches her, arguing with her for the first time, she dies of a heart attack.

Power and decline is the theme of many novels and stories about old and modern Cairo, including the realistic masterpiece that brought the author universal fame and the Nobel Prize, the Cairo Trilogy (*al-Thulathiyya*, 1956-1957):

Bayn al-qasrayn (1956), "Palace Walk"
Qasr al-shawq (1957), "Palace of Desire"
al-Sukkariya (1957), "Sugar Street"

The shadow of a dominant father, *al-Sayyid Abd al-Jawad*, falls over three generations of an ordinary Cairo family, whose fate and strivings we follow over the first part of the twentieth century. In this comprehensive family saga in three parts (the second of its kind in Egyptian literature after Taha Husayn's book about another family tree of misery, *Shajarat al-bu's*, 1944), different streams in contemporary Egypt are reflected in individual preferences and attempts at liberation among the siblings. These attempts include hedonism, intellectual life, communism as well as fundamentalist Islam.

The novels present a variety of outlooks from a local perspective, as does the work of Mahfouz as a whole. We get to know a genuine Cairo environment, are introduced to the private life both of an oppressed wife who is the source of love and care, and her closely guarded daughters. There are three very divergent sons, heading in opposite directions and in increasingly rebellious ways, as eventually the grandsons also do, though in an even more radical manner. The dignity or *karâma* of the father and the tradition, alongside the guidance of women, in a way holds the narrative as well as the family together. This is the basis of continuity as well as of opposition during the turbulent times of English occupation, and the constant confrontation of Occidental and Oriental values.

The Children of Gebelawi

In actual fact, a similar theme runs through *Awlad haratina* ("Children of our quarter", 1959; translated as *Children of Gebelawi*), which created the greatest stir in Mahfouz's life, and determined his fate in a violent way.

When the rich and powerful Gebelawi banishes his children he casts a spell on his family in doing so, as if he were expelling them from the

Garden of Eden. Patriarchal power determines their fate in this very complex narrative, where the incomprehensible main character has withdrawn from the scene and retreated to the house that can be seen at a distance. That is the closed core of all events, the invisible force in an expansive web of life.

In his novel, Mahfouz in a way writes a history of humankind up to the 1950s, starting with the Dawn of Creation. At the same time, it is a story about children in a Cairo suburb and their difficulties. For his narrative he has drawn on — or sometimes paraphrased — *Al-Qur'an* and the traditional Islamic *Hadith* literature for many figures and events, freely transforming them and inserting them into a new historical and completely fictional context. Leading characters in the novel were early identified as religious figures: Gebelawi (Jabalawi) himself has been identified as God (Allah), the Almighty Creator and Supreme Being. Similarly other main characters like Idris, Adham, Jabal, Rifa'a, Qasim, and 'Arafa have been seen as representing or symbolising Satan, Moses, the prophets Jesus and Muhammad, and modern science respectively. But the framework and the environment is a local area on the outskirts of Cairo at the foot of the Moqattam Hills, in a small-town atmosphere of family quarrels, hopes, and cares. Some inhabitants fear or follow local chiefs (politics?), whereas others turn to higher ideals (religion?).

When *Awlad haratina* was serialised in the Cairo newspaper *Al-Ahram* 1959, leaders of the Islamic university Al-Azhar, the custodian of faith and morals, called for the banishment of the 'heretical' book, and crowds of people marched on the streets to the big Al-Ahram building shouting their protests against the blasphemous book by Naguib Mahfouz. The ban was never officially sanctioned, the serialisation did not stop, but the novel was only published abroad in book form. In her well-informed dissertation, *The Limits of Freedom of Speech: Prose Literature and Prose Writers in Egypt under Nasser and Sadat* (Stockholm University, 1993), Marina Stagh showed the secret forces at work behind this public drama, and the final agreement reached between the author and the government in this affair. Today, there is no official ban on this book, and in fact, there never has been.

However, the publication of Salman Rushdie's *The Satanic Verses* in 1988, famous as well as infamous for its international consequences, raised the almost thirty-year old question of Mahfouz's blasphemy and his alleged undermining of the dignity of the Prophets. Suddenly Naguib Mahfouz found himself paired with a foreign author with whom he had nothing in common. Himself a pious Moslem believer, an author of international repute and learning and an earnest moralist, he felt obliged officially to defend Rushdie and the freedom of speech as a holy right of humanity. This made some of Rushdie's enemies compare *The Satanic Verses* to the *Children of Gebelawi*, concluding that a similar fatwa should have been pronounced on Mahfouz, too.

Naguib Mahfouz then summarised his moderate and very measured standpoint in this way:

> *"I have condemned Khomeini's fatwa to kill Salman*
> *Rushdie as a breach of international relations and as an*
> *assault on Islam as we know it in the era of apostasy. I*
> *believe that the wrong done by Khomeini towards Islam and*
> *the Muslims is no less than that done by the author himself.*
> *As regards freedom of expression, I have said that it must be*
> *considered sacred and that thought can only be corrected by*
> *counter-thought. During the debate, I supported the boycott*
> *of the book as a means of maintaining social peace, granted*
> *that such a decision would not be used as a pretext to*
> *constrain thought."*
> (*Al-Ahram*, 2 March, 1989; for a detailed account see Samia
> Mehrez, "Respected Sir", in: Michael Beard and Adnan
> Haydar, eds., *Naguib Mahfouz: From Regional Fame to*
> *Global Recognition*, Syracuse University Press, 1993.)

This did not finally settle the matter, however; nor could the stigma and the suspicion ever be effaced. The Egyptian Government offered the bigoted "infidel" (*kafir*) protection, but the author always refused, keeping to the simple routines of his private life in Cairo. One day in October 1994, however, on one of his regular visits to the Qasr Al Nil café, an attempt was made on his life by a follower of al-Jihad, the same religious group that assassinated Sadat. He was stabbed in his neck with a knife and was seriously injured, but survived.

The Enigmatic and the Absurd

It is easy, all too easy, to descry literary traces of the chaotic world of his life and times. There are many stories of modern decline and lack of order and confidence, but they are also literary experiments in modern prose genres.

In *Tharthara fawq al-Nil* (translated as *Adrift on the Nile*), 1966, a complete opposite to Meriamun's floating along the Nile in *Dweller in Truth*, we are on a house-boat on the Nile among disillusioned and cynical gamblers, loafers and addicts. The government official, Anis Zaki, who has lost his wife and his daughters, is the centre of a whirl of frustrated talk and passivity. He and his friends are involved in a fatal accident where a peasant is killed, but they let the matter rest there.

In the Alexandria novel *Miramar* (1967), set in the early 1950s, a maid, Zahra, attracts the attention of a number of men in the neighbourhood. Different narrators tell different stories in this little drama, which mirrors the various current political and social views of the revolutionary years. The author shows a complex setting, but does not supply us with a key to show us how to interpret the set of narratives.

Taht al-Mizalla ("Under the [bus] Shelter") is a collection of short stories from 1969. In the title story, we are presented with a crowd of people at a bus stop, waiting, watching. The most horrifying and upsetting things happen before their eyes while they are standing there, odd occurrences and violent accidents to which they are passive witnesses. We do not know what they think, nor do we know the meaning of the scene. It is like a late surrealistic film by Luis Buñuel or an absurd drama by Eugène Ionesco. In *Hadrat al-Muhtaram*, "Respected Sir" (1975), Mahfouz satirically portrays a civil servant and government employee, 'Utman Bayyumi, in which he most probably also ironically portrays himself as an aspiring official, a part of the system and the class he satirises.

The Court of Osiris

The author's concern with power is obvious, but his main question is the following: What have the leaders of Egypt — pharaohs, sultans, khedives, kings, presidents, fathers, invaders, occupants — done for (or done to) his country? In this comparative study of politics, focused both on the past and on the present, parallels and repetitions stand out, and the observer seems to behold an eternal return of decline and fall, nationally as well as individually. He does not hesitate to let the high and the mighty confront each other, and there are some extreme and highly controversial examples of this feature.

In *Amam al-'Arsh* ("Before the Throne"), published in 1983, Egyptian leaders of different eras are assembled and committed for trial at a court. The justice in this Supreme Court is the sun god Osiris, sitting on his throne. The goddess Isis and their son, falcon-headed Horus, assisting him, sits by his side. It is like a scene from the ancient Egyptian *The Book of the Dead*, written in hieroglyphs. That is the manual for the dead, a guide-book for those who are about to be tried by the eternal court, a fragmentary "book" that has been found in scrolls and mural paintings in tombs throughout the country.

Now the court of Osiris is going to judge the rulers of Egypt, from the founding father of the country, Menes (the Pharaoh Narmer, who united Upper and Lower Egypt five thousand years ago) through the Ottoman rulers down to Anwar el-Sadat, killed in 1981, shortly before the novel was written. Pre-Arabic history thus blends and intertwines with Islamic history and contemporary politics. The author perceives, or tries to establish, a deeply felt continuity. Sticking to his identity as an inhabitant of Old Cairo, al-Qahirah, he has also publicly objected to various kinds of pan-Arabic movements in the Muslim world, stressing that the Al-Azhar mosque and university in his vicinity is the centre of the teaching of *sunnah*. Being the child of two civilisations, the Egyptian civilisation is as important to him as the Islamic one.

In the crowded court room where the voices of poets, khedives, Sufi women, patriarchs, kings, pharaohs, presidents, courtiers and many others are heard, examination and interrogation sometimes turn into heated accusations and debates, since the defendants partake in the trial of the accused. From the 19th dynasty (1308–1186 BC), a period of military campaigns against Hittites and others, strong domination, and the building of great temples in Abu Simbel, Karnak, Luxor and Thebes, the Pharaoh Rameses II appears. After being questioned, he compliments his late successor, the revolutionary Gamal Abd el-Nasser.

Nasser seized power in 1956 after the overthrow of King Faruq in 1952 and the abolition of the monarchy in 1953, when the Republic of Egypt was proclaimed. The ancient king's admiration of the republican may in part be the author's assessment of Nasser, emphasising parallels between the two characters. Judge Osiris furthermore points out that President Nasser was the first ruler to care for his people. But Rameses II, the ruler of an empire, blames Nasser for reducing Egypt to an insignificant state, and the founder of Egypt, the Pharaoh Menes, chimes in accusing Nasser of having let the features of majestic Egypt dissolve into the vague outlines of Arabism.

Nasser, in his turn, attacks his successor, Anwar el-Sadat, for having destroyed the country by his open door-policy, the *infitah*, "opening up" of Egypt, leaving the door ajar for capitalism, American influence, and corruption. Sadat defends himself very ably, but Nasser's reproaches are sharp and to the point. It is almost impossible to settle the proportion of inequity between them: in the course of the dialogue, both are harshly criticised. However, Sadat gets unexpected support from one of the most controversial among the famous pharaohs — Akhenaten, the "heretical" pharaoh, Amenophis IV.

Whereas the warrior-king Rameses II partly recognised himself in Nasser, the first president of Egypt, a peace-making predecessor of the 18th dynasty thus sides with the president's successor and heir. Akhenaten meddles in the dispute, stating that Sadat aimed at peace, that he worked for peace as he, Akhenaten, had done in his day, and that they were both unjustly denounced for faithlessness. The monotheist Akhenaten thus identifies with Sadat. But who was Akhenaten, then? The question seems to be of central importance to the author.

Who Was Akhenaten?

Two years later, Naguib Mahfouz published one of his most intriguing and perhaps most revealing books; his novel on Akhenaten, *al-'A'ish fi-l-haqiqa* (1985), "Dweller in Truth".

Adopting the narrative method of multiple narrators, the novelist approaches a well-known enigma, the identity of Akhenaten, the author of beautiful hymns to the sun and the predecessor of Tutankhamun. King Amenophis IV, who ruled between about 1375 and 1358 BC, abandoned the dynastic cult of Amun and other gods in Thebes and left the capital, aiming to build a new one in the place now known as Tell El-Amarna. He devoted all his worship to the solar disc of Aten, the sun god, represented by rays that terminate in hands holding the signs of *ankh*, "the force of life". In a similar fashion, the famous Egyptian obelisks are holy representations of sunrays.

Aten, a form of the ancient solar god Re or Ra, was hailed as the almighty, universal god. The king himself changed his name into Akhenaten, "he who is useful to Aten", and he called his new city Akhet-Aten, the "horizon of Aten". During his reign, Egypt was weakened by his revolutionary religious and political reforms and finally the non-belligerent dissenter was overthrown. His successor Tutankhamun (Tut-Ankh-Amun) abandoned all the reforms and reinstated Thebes as the capital. The city of Akhet-Aten was deserted and fell into decay.

At that point in history, Mahfouz's novel begins. Some years after the fall of Akhenaten, the protagonist Meriamun (whose name deceptively reminds us of Meritamun's mummy in the Egyptian Museum of Cairo) and his father are sailing along the Nile on a voyage from their home in Sais during the inundation. Suddenly they pass a strange city. Meriamun observes that the roads are empty, that the trees have no leaves, and that all the gates and windows are closed like the eyelids of the dead. It is completely silent. From his father he learns that this is the city of the heretical and faithless pharaoh. It is not completely deserted, though. The former wife of the deceased apostate is still living there. The foremost among Egyptian beauties, Queen Nefertiti, is living alone in the empty city, as though in a jail. The name Nefertiti means "the beautiful (one) is come".

Meriamun then recalls what he has heard about the whirlwind that devastated the Egyptian empire, the era called the War of the Gods, and the stories about the young pharaoh who revolted against his father, challenged the priests and fate itself. Meriamun's curiosity grows into a passion to know the truth about the strange king. His prominent father assents, himself being "a dweller in truth". But unlike his father, the boy has another mission: not rooted in "truth" or in any conviction as his father's was. He starts his pursuit of knowledge, eager to find out on his own. He acts like a modern investigator, a scholar and a sceptic, a believer in science and free inquiry. It is easy, perhaps too easy, to regard him as an alter ego of the author, and his perennial pursuit of truth as a guideline to the works of Mahfouz. Nevertheless, just as Meriamun becomes eager to learn about Akhenaten, so do we become eager to know the fictional Meriamun.

Meriamun goes to Thebes to see people who knew the king, and finally he visits Nefertiti in the fallen city. This way fourteen different stories are told, and equally many versions of the truth. He interviews Akhenaten's high priest Meri-Ra, and then we observe that Meriamun is himself named after the reinstalled god Amun. He asks the king's mother Tiye about her son, he interrogates his secretary, the chief of the police, Nefertiti's father, and so forth. Meriamun (or the reader) is left none the wiser. Or rather: the conclusion is that it is up to the reader to determine the truth about Akhenaten's character, his regime and his times. Alternatively, truth is pictured as unattainable, invisible, evasive, like the person called Akhenaten and the god called Aten. These features of the god and the king are those that stand out in the variety of accounts, and perhaps one more: a dislike of warmongering.

Nefertiti, the last person interviewed, does not really know either, but says her former husband was assassinated. (He thus met with the same fate as Anwar el-Sadat, the late pharaoh the king identified with in the novel *Before the Throne*.) She looks forward to see her beloved in the next life, reuniting with him forever, and she hopes to sit at his dear side at the Throne of Truth. Akhenaten will thus be exalted as Osiris, the Supreme Judge in *Before the Throne.*

As Meriamun observes, the way to truth in this world is without beginning and end, since those who are drawn to eternal truth will always extend it. Consequently, history is without beginning or end, like the existential *Hikaya bi-la bidaya wa la nihaya*, the 1971 story without beginning or end.

Meriamun afterwards tells everyone he knows about his investigations, and what various people have told him, but two things he discloses only to us: his increasing admiration of the hymns to the one and only God, and his love of the beautiful Nefertiti. In these concluding words of the novel, one may sense a combined veneration of literary and religious values, deference to the sublime, the invisible and unattainable, be it Allah (the one God) or the truth of humankind or that of any single individual. They all harbour the mystery of Aten and Akhenaten.

Like Osiris, Mahfouz examines and assesses the standards of Egyptian rule, life and manners. Just as at the court of Osiris, Isis, and Horus, there are a number of questions and responses, but we are left without answers. The author's quest is a pursuit without end.

❄

At the bus-stop shelter, walking through the ages.
Photo courtesy of Mohamed Hegazy.

Patrick WHITE
Photo courtesy of the Nobel Foundation.

Patrick White – Autobiography

✳

I was born on May 28th 1912 in Knightsbridge, London, to Australian parents. Victor White was then forty-two, his wife, Ruth Withycombe, ten years younger. When I was six months old my parents returned to Australia and settled in Sydney, principally because my mother could not face the prospect of too many sisters-in-law on the property, in which my father had an interest, with three older brothers. Both my father's and my mother's families were yeoman-farmer stock from Somerset, England. My great-grandfather White had emigrated to New South Wales in 1826, as a flockmaster, and received a grant of crown land in the Upper Hunter Valley. None of my ancestors was distinguished enough to be remembered, though there is a pleasing legend that a Withycombe was fool to Edward II. My Withycombe grandfather emigated later in the nineteenth century. After his marriage with an Australian, he and my grandmother sailed for England, but returned when my mother was a year old. Grandfather Withycombe seems to have found difficulty in settling; he drifted from one property to another, finally dying near Muswellbrook on the Upper Hunter. My father and mother were second cousins, though they did not meet till shortly before their marriage. The Withycombes enjoyed less material success than the Whites, which perhaps accounted for my mother's sense of her own superiority in White circles. Almost all the Whites remained wedded to the land, and there was something peculiar, even shocking, about any member of the family who left it. To become any kind of artist would have been unthinkable. Like everybody else I was intended for the land, though, vaguely, I knew this was not to be.

My childhood was a sickly one. It was found that I was suffering from nothing worse than asthma, but even so, nobody would insure my life. As a result of the asthma I was sent to school in the country, and only visited Sydney for brief, violently asthmatic sojourns on my way to a house we owned in the Blue Mountains. Probably induced by the asthma, I started reading and writing early on, my literary efforts from the age of about nine running chiefly to poetry and plays. When thirteen I was uprooted from Australia and put at school at Cheltenham, England, as my mother was of the opinion that what is English is best, and my father, though a chauvinistic Australian, respected most of her caprices. After seeing me 'settled' in my English prison, my parents and sister left for Australia. In spite of holidays when I was free to visit London theatres and explore the countryside, I spent four very miserable years as a colonial at an English school. My parents returned for the long holiday when I was sixteen, and there were travels in Europe, including Scandinavia. Norway and Sweden made a particular impression on me as I had discovered Ibsen and Strindberg in my early teens — a taste my English housemaster deplored: "You have a morbid kink I mean to stamp out"; and he then proceeded to stamp it deeper in.

When I was rising eighteen I persuaded my parents to let me return to Australia and at least see whether I could adapt myself to life on the land before going up to Cambridge. For two years I worked as jackeroo, first in the mountainous southern New South Wales, which became for me the bleakest place on earth, then on the property of a Withycombe uncle in the flat, blistering north, plagued alternately by drought and flood. I can remember swimming my horse through floodwaters to fetch the mail, and enjoying a dish of stewed nettles during a dearth of vegetables. The life in itself was not uncongenial, but the talk was endlessly of wool and weather. I developed the habit of writing novels behind a closed door, or at my uncle's, on the dining table. More reprehensible still, after being a colonial at my English school, I was now a 'Pom' in the ears of my fellow countrymen. I hardly dared open my mouth, and welcomed the opportunity of escaping to King's College, Cambridge. Even if a university should turn out to be another version of a school, I had decided I could lose myself afterwards as an anonymous particle of the London I already loved.

In fact I enjoyed every minute of my life at King's, especially the discovery of French and German literature. Each vacation I visited either France or Germany to improve my languages. I wrote fitfully, bad plays, worse poetry. Then, after taking my degree, the decision had to be made: what to do? It was embarrassing to announce that I meant to stay in London and become a writer when I had next to nothing to show. To my surprise, my bewildered father, who read little beyond newspapers and stud-books, and to whom I could never say a word if we found ourselves stranded alone in a room, agreed to let me have a small allowance on which to live while trying to write.

At this period of my life I was in love with the theatre and was in and out of it three or four nights of the week. I tried unsuccessfully to get work behind the scenes. I continued writing the bad plays which fortunately nobody would produce, just as no one did me the unkindness of publishing my early novels. A few sketches and lyrics appeared in topical revues, a few poems were printed in literary magazines. Then, early in 1939, a novel I had managed to finish, called *Happy Valley*, was published in London, due to the fact that Geoffrey Grigson, the poet, then editor of the magazine *New Verse* which had accepted one of my poems, was also reader for a publishing firm. This novel, although derivative and in many ways inconsiderably, was received well enough by the critics to make me feel I had become a writer. I left for New York expecting to repeat my success, only to be turned down by almost every publisher in that city, till the Viking Press, my American publishers of a lifetime, thought of taking me on.

This exhilarating personal situation was somewhat spoilt by the outbreak of war. During the early, comparatively uneventful months

I hovered between London and New York writing too hurriedly a second novel, *The Living and the Dead*. In 1940 I was commissioned as an air force intelligence officer in spite of complete ignorance of what I was supposed to do. After a few hair-raising weeks amongst the RAF greats at Fighter Command I was sent zigzagging from Greenland to the Azores in a Liverpool cargo boat with a gaggle of equally raw intelligence officers, till finally we landed on the Gold Coast, to be flown by exotic stages to Cairo, in an aeroplane out of Jules Verne.

The part I played in the war was a pretty insignificant one. My work as an operational intelligence officer was at most useful. Much of the time was spent advancing or retreating across deserts, sitting waiting in dust-ridden tents, or again in that other desert, a headquarters. At least I saw something of almost every country in the Middle East. Occasionally, during those years bombs or gunfire created what should have been a reality, but which in fact made reality seem more remote. I was unable to write, and this finally became the explanation of my state of mind: my flawed self has only ever felt intensely alive in the fictions I create.

Perhaps the most important moments of my war were when, in the western desert of Egypt, I conceived the idea of one day writing a novel about a megalomaniac German, probably an explorer in nineteenth century Australia, and when I met my Greek friend, Manoly Lascaris, who has remained the mainstay of my life and work.

After demobilisation we decided to come to Australia where we bought a farm at Castle Hill outside Sydney. During the war I had thought with longing of the Australian landscape. This, and the graveyard of postwar London, and the ignoble desire to fill my belly, drove me to burn my European bridges. In the meantime, in London, in Alexandria on the way out, and on the decks of liners, I was writing *The Aunt's Story*. It was exhilarating to be free to express myself again, but nobody engaged in sorting themselves out of the rubble left by a world war could take much interest in novels. Australians, who were less involved, were also less concerned. Most of them found the book unreadable, just as our speech was unintelligible during those first years at Castle Hill. I had never felt such a foreigner. The failure of

The Aunt's Story and the need to learn a language afresh made me wonder whether I should ever write another word. Our efforts at farming — growing fruit, vegetables, flowers, breeding dogs and goats, were amateurish, but consuming. The hollow in which we lived, or perhaps the pollen from the paspalum which was always threatening to engulf us, or the suspicion that my life had taken a wrong turning, encouraged the worst attacks of asthma I had so far experienced. In the eighteen years we spent at Castle Hill, enslaved more than anything by the trees we had planted, I was in and out of hospitals. Then about 1951 I began writing again, painfully, a novel I called in the beginning *A Life Sentence on Earth*, but which developed into *The Tree of Man*. Well received in England and the United States, it was greeted with cries of scorn and incredulity in Australia that somebody, at best, a dubious Australian, should flout the naturalistic tradition, or worse, that a member of the grazier class should aspire to a calling which was the prerogative of schoolteachers! *Voss*, which followed, fared no better: it was 'mystical, ambiguous, obscure'; a newspaper printed its review under the headline 'Australia's most Unreadable Novelist'. In *Riders in the Chariot* it was the scene in which Himmelfarb, the Jewish refugee, is subjected to a mock crucifixion by drunken workmates which outraged the blokes and the bluestockings alike. Naturally, 'it couldn't happen here' — except that it does, in all quarters, in many infinitely humiliating ways, as I, a foreigner in my own country, learned from personal experience.

A number of Australians, however, discovered they were able to read a reprint of *The Aunt's Story*, a book which had baffled them when first published after the war, and by the time *The Solid Mandala* appeared, it was realised I might be something they had to put up with.

In 1964, submerged by the suburbs reaching farther into the country, we left Castle Hill, and moved into the centre of the city. Looking back, I must also have had an unconscious desire to bring my life full circle by returning to the scenes of my childhood, as well as the conscious wish to extend my range by writing about more sophisticated Australians, as I have done in *The Vivisector* and *The Eye of the Storm*. On the edge of Centennial Park, an idyllic landscape surrounded by a metropolis, I have had the best of both worlds. I have tried to celebrate the park, which means so much to so many of us, in *The Eye of the Storm* and in some of the shorter novels of *The Cockatoos*. Here I hope to continue living, and while I still have the strength, to people the Australian emptiness in the only way I am able.

✳

Together with his friend Manoly Lascaris, Patrick White moved to this house at Centennial Park, Sydney, in 1965. The house figures in the novel *The Eye of the Storm*.

Photo courtesy of Karin Hansson.

"The ultimate paradox is his pessimism as to the possibility of communicating what any single individual can perceive."
Patrick White in Sydney in 1973.
Photo by Ingmar Björkstén. Courtesy of the Bonnier Archives and the Nobel Foundation.

Patrick White – Existential Explorer

❄

Karin Hansson

The Nobel Prize

When Patrick White was awarded the Nobel Prize in Literature in 1973, the Swedish Academy's commendation referred to the author's epic and psychological narrative art as having introduced a new continent into literature. This standpoint may seem surprising now, but at that time it was a generally valid Swedish (and European) perspective — up to then literary criticism had largely ignored post-colonial writing and other new literatures in English. In many non-European countries, however, Patrick White was a well-known name, and he had already won prestigious prizes. His Nobel Literature Prize was the first to be awarded to an Australian, it is true, but the quality of Australian literature in general and of Patrick White's writing in particular had long been recognized, not only by Australians, but also outside the country. To White's fellow-countrymen, the Nobel Prize confirmed his status as a major novelist whose fiction had, for more than twenty years, been regarded as an important contribution to the literature of the English-speaking world.

With the prize-money he created the PATRICK WHITE LITERARY AWARD to encourage the development of Australian literature. The committee has been instructed to give precedence to authors whose writing has not yet received due recognition. Among the winners have been Christina Stead, Randolph Stow, Thea Astley, and Gerald Murnane.

European or Australian?

It is important to recognize White's conflicting loyalties to Europe and Australia. In many respects the European background and influences are obvious in his writing. After spending his first school years at private schools in New South Wales, he was sent at the age of thirteen, very much against his will, to Cheltenham College in England. There he spent a "four-year prison sentence," according to his autobiography, *Flaws in the Glass* (1981). All the same, after a couple of years as a jackeroo on his uncle's sheep station in Australia, he chose to return to England and Cambridge in order to study modern languages. He also spent some vacations in Germany and developed a lasting interest in German and French literature.

After taking his degree, intent on forging a career for himself as a writer, he settled in London to write novels, plays and poetry. His first novel, *Happy Valley* (1939) was praised by some British critics, but the next, *The Living and the Dead* (1941), forced out prematurely because of the impending war, was a failure even according to the writer himself. The same year he received his Royal Air Force posting as an Intelligence Officer in the Middle East and Greece. His experiences in the Western Desert led him to the reading of Australian explorers. Eire's *Journal of Expeditions into Central Australia* evoked in him "the terrible nostalgia for the desert landscapes," a feeling that was to influence his later novels, above all *Voss*. His first visit to his native country after the war made him decide to settle in Australia. He was then nearly forty years old.

Even though all the later novels are wholly or mainly set in Australia, they belong to the European epic tradition insofar as they are inspired by and based on Greek mythology, Judeo-Christian mysticism, C. G. Jung's psychology, and the Joycean stream-of-consciousness technique. White has quite often been compared to the greatest among Russian and French novelists. After his return, external Australian conditions and details merged with abstractions from his European experience. In consequence, tensions both between Europe and Australia and between a "real" and a "symbolic" Australia became significant. There is also an obvious love-hate relationship with the country of his childhood. His disappointment in the materialism and shallowness of what he terms the Great Australian Emptiness, is very marked in the essay "The Prodigal Son" (1958), in which he expresses serious misgivings about the country's future. This attitude also comes to the fore in his writing, and caused many critics to accuse him of elitism and intellectual snobbery. White's "Australianness" and commitment to the continent are nevertheless generally recognized in the sense that his return brought true colours back to his palette and, in *The Aunt's Story* (1947), introduced a new style into his canon, initiating novels of depth and dedication.

The First Phase

The Aunt's Story and *The Tree of Man*

For many reasons, the two novels *The Aunt's Story* and *The Tree of Man* can be considered as the initial phase of a novel sequence. Unlike the earlier books, they are concerned with the most fundamental issues of humanity, such as the relationship between madness and sanity, reality and illusion, and the problem of communication in existential matters. For the first time, the idea of movement is structurally and consistently combined with the search for ultimate truth, the quest. Here, as in his later books, the characters are divided into two categories from a spiritual point of view: the living and the dead. The true questers are explicitly heading for salvation or damnation in a religious sense, not merely for an intenser form of life. They are opposed to the materialistic characters, who are rooted in social conventions, dreaming of possessions and gentility.

The Aunt's Story presents the quest of Theodora Goodman, a spinster, "dry, leathery and yellow." She is the first in a series of alienated, humble seekers in search of a true reality behind the appearances of social life. Like Theodora, they are all facing the choice "between the reality of illusion and the illusion of reality." Increasingly estranged from ordinary life, she passes through a series of heightened moments of insight and awareness, "epiphanies" to use James Joyce's term, leading up to final illumination. The question is asked whether spiritual illumination is compatible with the retention of selfhood and mental sanity. Theodora Goodman's ending is characteristically ambiguous. To the world, she appears to be mentally deranged when she is finally led away by Doctor Holstius, whose name hints at a sense of ultimate wholeness. Holstius knows that "lucidity isn't necessarily a perpetual ailment," and he summarises the lesson she has learnt: "You cannot reconcile joy and sorrow. Or flesh and marble, or illusion and reality, or life and death. For this reason, Theodora Goodman, you must accept." Like all the later questers, she experiences that "for the pure abstract pleasure of knowing, there was a price paid."

The changes in setting between the various parts of the novel are related to the mental development of the protagonist to such an extent that it is unclear whether Theodora's odyssey carries her across oceans and continents in a physical sense, or whether the whole journey takes place in her troubled mind. There is a significant epigraph from Olive Schreiner for the third part of the novel: "When your life is most real, to me you are mad."

The Tree of Man was the first novel to win international acclaim. On the narrative level, it deals with the hardships of a farming couple, Stan and Amy Parker, in New South Wales. The frame is that of the conventional pioneering saga, albeit with biblical overtones and associations. Stan is one of White's characteristic seekers, and his spiritual capacity is set off against Amy's conventional attitudes. Stan "respected and accepted her mysteries, as she could never respect and accept his." The story contains the typical features of Australia's natural trials and disasters, such as bushfires, drought, and floods, but above all it enacts the psychological drama of Stan's desire to understand the purposes of God, which "are made clear to some old women, and nuns and idiots." The background and aim of the novel are indicated in White's essay "The Prodigal Son":

> *It was the exaltation of the "average" that made me panic*
> *most, and in this frame of mind, in spite of myself, I began*
> *to conceive another novel. Because the void I had to fill was*
> *so immense, I wanted to try to suggest in this book every*
> *possible aspect of life, through the lives of an ordinary man*
> *and woman. But at the same time I wanted to discover the*
> *extraordinary behind the ordinary, the mystery and the*
> *poetry, which alone could make bearable the lives of such*
> *people, and incidentally, my own life since my return. So*
> *I began to write* THE TREE OF MAN.

Like Theodora, just before he dies Stan is granted a moment of illumination and universal harmony, finally understanding that "One, and no other figure is the answer to all sums."

The Second Phase

Voss, Riders in the Chariot, and *The Solid Mandala*

If the novels of the first phase focus on one single protagonist, those of the second are built up around more than one central character; two, four, and two respectively. All of them are presented as being exceptional, different, puzzling, or even repulsive. In this respect, White's theme is the reverse, namely to show the ordinariness in which man's divinity is contained behind outward exceptionality. The characters are also conscious of their otherness in a way Theodora and Stan were not, and they seem to accept it to the extent that they even seem to take a pride in it. All three books have a dimension of religious mysticism and contain celestial images related to different myths and archetypes: the Comet, the Chariot, the Sun and the Moon. The mandala belongs to the same sphere. In this structural system we find symmetrically grouped principal figures such as the circle, the square, and the cross. In all three novels, there is a general movement towards a centre representing wholeness, psychological and religious.

We also find elements from the mythological beliefs of the Australian aborigines.

Voss is based on the story of the German explorer Ludwig Leichhardt. Like him, Voss intends to cross the Australian continent from east to west with a group of men. The three stages of his quest are represented by coast/city, bush, and interior/desert, respectively. Voss, dominated by willpower, megalomania and egocentricity, has been compared both to Faust and Hitler. The desert, he thinks, will prove a worthy opponent when he wants to demonstrate his superhuman status. Before he sets out he meets Laura Trevelyan, a spinster who makes a strong impression on him. During his journey into the desert he becomes transformed through her telepathic influence as they communicate in dreams and delirious visions. In the end, he is humbled through Christ-like suffering and death. Laura believes that his union with the desert of the interior implies a different kind of victory: "Voss did not die." He is still there, it is said, in the country, and always will be. "His legend will be written down, eventually, by those who have been troubled by it." Through their shared experience, she too has learnt that "Knowledge was never a matter of geography. Quite the reverse, it overflows all maps that exist. Perhaps true knowledge only comes of death by torture in the country of the mind."

The epigraph of *Riders in the Chariot* is taken from William Blake and suggests connections with Ezekiel and Isaiah, the Chariot itself representing God as divine grace as well as destructive terror and judgement. The corresponding Christ-figure in the novel is Himmelfarb, a Jewish refugee from Germany. Like Voss, he is associated with the element of fire. The other illuminati are equally insignificant from a social point of view: Mary Hare, an elderly spinster; Ruth Godbold, a poor and hard-working housewife; and Alf Dubbo, a part-Aboriginal painter. Himmelfarb's death by crucifixion on Good Friday suggests that the capacity for evil in Australian suburbia is comparable to its realisation in Nazi Germany as well as in the Bible. Alf Dubbo expresses the unity of the four illuminati in his last painting, which represents the chariot of fire. "Their hands, which he painted open, had surrendered their suffering, but not yet received beatitude."

The central characters in *The Solid Mandala* are the elderly twins Arthur and Waldo Brown. Just like Voss and Laura, they are contrary beings, or rather contrasting aspects of the same being, contending but wholly interdependent. In *Flaws in the Glass*, White claims that the Brown brothers represent his two halves: "Waldo is myself at my coldest and worst." Mentally retarded Arthur, a Christ-figure like Himmelfarb, assumes the burden of suffering, responsibility and guilt to rescue that other part of himself, and when Waldo dies, Arthur proclaims himself guilty and ends up in an asylum.

The mystery of failure is a common denominator in the novels of the second, religious, phase. As stated in *Voss*: "The mystery of life is not solved by success, which is an end in itself, but in failure, in perpetual struggle, in becoming". Similarly, *Riders in the Chariot* suggests that "atonement is possible perhaps only where there has been failure."

The Third Phase

The Vivisector and *The Eye of the Storm*

The fact that almost five years elapsed between *The Solid Mandala* and *The Vivisector* indicates a change of emphasis. A single protagonist is at the centre of both novels in the third phase, and they are dominated by the image of the Eye, which is given a multidimensional function. The central character in *The Vivisector* is Hurtle Duffield, a painter whose artist's eye indicates his special instinct, enabling him to discern truth behind appearances. As the title suggests, the eye is also a knife, an instrument of torture. The artist's fellow-beings must try to protect themselves from "his third eye," used ruthlessly to vivisect vulgarity, pretentiousness and falsity.

One of his paintings represents the Mad Eye, signifying God as artist, vivisector and opponent. The dual aspect, creation and destruction, the artist as torturer and victim, divine truth as the goal of human will, and human will as subordinate to inevitability, pervades the whole novel. Causing his victims to suffer through his vivisecting eye, the artist must also suffer, but to an even greater extent. Before he reaches illumination, typically through humiliation, his partiality to comparing the artist to God corresponds to Voss's blasphemous belief in his own divine capacity. Like the explorer, the artist approaches divinity only by becoming human.

In *The Eye of the Storm*, the central character is Elizabeth Hunter, a dying old woman who is almost blind. She is greedy and cruel, and all her life she has consumed others. The title refers to a climactic moment in her life when she was left alone on an island and was caught in a tropical cyclone. The experience implied suffering and humiliation, closeness to death, but also a moment of incredible grace and stillness. Her dying implies a renewed search for that moment of grace, and her deathbed becomes the still centre in all the emotional storms that surround her. To her, as to Hurtle Duffield, the eye comes to stand for the core of reality, the centre of our true existence inside all the layers of appearance. In the end, both protagonists become obedient instruments of the Divine Eye. Their will is wholly concentrated on reaching the eye of truth and infinity, a process that ultimately implies the destruction of that same will. They both make the act of dying a work of art, and the two novels end on a positive note by combining the fundamentally human aspect with the concept of divinity.

The Last Phase

A Fringe of Leaves and *The Twyborn Affair*

The two books representing the last phase deal with the efforts of the central characters to achieve self-discovery. But neither arrives at the kind of illumination that was typical of the earlier books. The religious dimension is toned down in favour of social and psychological ones, and the imagery is less complex. *A Fringe of Leaves* is a historical novel in the sense that the story of its heroine, Ellen Roxburgh, is based on that of Eliza Fraser, who in 1836 was shipwrecked on a reef off the Queensland coast. She was the only survivor and spent six months, totally naked, with the aboriginal tribe that saved her. In the novel, the idea of nakedness suggested by the title is related to the general message. To the very last, Ellen Roxburgh defends her fringe of leaves, her "gesture to propriety." She does not, like Voss, become part of the Australian continent in a dying vision of the Southern Cross. Nor does she experience the blessing of sea, sky and land in the way Elizabeth Hunter did. Her return to England suggests that she is withdrawing, physically and spiritually, that she lacks the strength to go to such extremes as White's true questers do.

In the earlier novels the themes of homosexuality and transvestism are touched upon only incidentally. In *The Twyborn Affair* they come into focus for the first time. The central character, Eddie/Eudoxia/Eadith Twyborn, has a male body and a female consciousness, and in his search for identity he uses various external disguises, which keep even the reader in the dark. In one section he appears as the young wartime hero, in another he/she assumes the part of Mrs. Trist, the keeper of a fashionable London brothel. He fails to find fulfilment and true sexual identity either as a man or as a woman, but learns to value friendship and the importance of recognising the woman in man and the man in woman. The novel can be read as an enquiry into bisexuality, and sees androgyny as a symbol of wholeness.

An Existential Pessimist

Taken together, Patrick White's novels express no specific orthodoxy or conviction concerning existential, mystical or psychological matters, even though it is obvious that he has been inspired by Judeo-Christian mysticism and the philosophies of Eckhart, Schopenhauer, Jung, Buber, and Blake among others. In *Voss*, for instance, there is a fusion between the more traditional Christian and Dantean aspects, the mystical writings of Meister Eckhart, and the mythological beliefs of the Australian aborigines in a way that is typical of White's literary technique. In other books, too, we find layers of significance in which White uses various Australian associations combined with archetypal or literary European ones. His attitude is pessimistic in the sense that his successful questers are described as innocent simpletons, isolated or alienated, physically or mentally handicapped, whose final insight is achieved only through ordeals and humiliation. As figures like Voss and Hurtle Duffield show, the explorer/artist is necessarily a sufferer who, in his search for truth, causes others to suffer too. White, like his protagonists, asks himself: "Am I a destroyer, this face in the glass, which has spent a lifetime searching for what it believes, but can never prove, the truth? A face consumed by wondering whether truth can be the worst destroyer of all."

Australia as the country of the mind exemplifies White's belief that "the state of simplicity and humanity is the only desirable one for artist or for man. While to reach it may be impossible, to attempt to do so is imperative." In his vocabulary "knowledge" is always unintellectual, irrational and intuitive. It refers to things you "do not know, but know", to use a wording which, with slight variations, occurs in all his books. He is scornful and sarcastic about the complacent attitudes in those who are satisfied with spiritually limited horizons, those who show no interest in "the mystery and poetry" beyond the Great Australian Emptiness. All his novels demonstrate how will-power, permanence, possessions and safety must be sacrificed. A journey of exploration is always a painful and solitary business through "that solitary land of the individual experience, in which no fellow footfall is ever heard", as it says in the first epigraph in *The Aunt's Story*.

In this division into "the living and the dead", we are given to understand that there is no rational explanation of why some are granted the experience of unity and illumination, although it becomes obvious that they have a special relationship with the landscape. Thus the desert is said to belong to Voss merely "by right of vision." In many contexts there is a logical but unresolved tension between the spiritual seeker and the surrounding society. The recurring notions of suffering and loss of self-will indicate that the state in which his questers live is not compatible with an ordinary social existence. Without exception they are described as socially inhibited because of their otherness and the power of their individual vision. An effect of paradoxical elitism is achieved when their outward inferiority and spiritual superiority is portrayed against the background of mental mediocrity and materialism.

Voss's (and White's) Australia is a country of infinite distances, unimaginable age and mystery, and the grandeur of this mental landscape is set off against the complacent mediocrity of the masses huddled in the coastal cities. The reason why Patrick White has frequently been accused of elitism seems to be his fierce disdain of the commonplace, his horror of the average, and his contempt for the materialist attitudes of mundane suburbia that are manifest in all his writing. The desert not only assumes the unexpected and unusual nostalgic quality previously mentioned, but at the same time retains its archetypal, apocalyptic, and daemonic properties. Traditionally a place where deep truths are revealed, it is at the same time a place of suffering and hardship. The enormous distances are used to illustrate the discrepancies between "aspiration and human nature" or between "earthbound flesh and aspiring spirit."

The desert in the centre of the Australian continent, conveniently shaped like a human heart, becomes the Interior with all its interpretative possibilities. The notion of mental exploration is linked with geographical imagery, particularly that of the desert/wilderness. Theodora Goodman becomes a timeless explorer in the wake of Odysseus, and the fact that Stan Parker is named after Stanley, the explorer, becomes significant in the context. Voss travels into the desert in Leichhardt's footsteps and Himmelfarb, too, becomes a desert explorer in his last vision of a journey where "whole deserts were crossed." Towards the end of his life Hurtle Duffield, the painter, has to reassess what had up till then been the basis o his credo, and to face "the vastest desert he had ever set out to cross: not the faintest mirage to offer illusory solace." Even the old woman's death-bed in *The Eye of the Storm* is a symbolic desert in which she spent years "lying on this mattress of warm moist sand." Finally her desert, too, takes on its apocalyptic aspects, and she is granted "some miraculous dispensation to feel sand benign and soft between the toes." At last she herself becomes part of "this endlessness." Himmelfarb's death takes place in Sarsaparilla, White's hellish suburbia, which represents another kind of desert, "the Great Australian Emptiness, in which the mind is the least of possessions", the recurrent target of White's satire.

To be able to appreciate Patrick White's consistent but complex structures and imagery, the reader must be prepared to enter his symbolic Australia with the open-minded attitude of the explorer or the child. He or she must not demand verisimilitude or logical argumentation, as this is a territory that is usually reserved for poetry. The author's methods resemble those of a poet, musician or painter. In spite of the fact that White has an idealistic and passionate belief in the power of art and literature to change the world, he claims that he writes only for himself and has "never thought about readership." The ultimate paradox offered by this paradoxical writer is his pessimism as to the possibility of communicating what any single individual can perceive within his own experience. To thousands of readers, his writing is an entity which consistently explores and communicates his perception of reality. For them, it contradicts the view he expressed above.

❋

Uluru (Ayers Rock), the most famous landmark of the Australian desert landscape.
Photo courtesy of Karin Hansson.

Ernest HEMINGWAY

Photo courtesy of the Nobel Foundation.

A Case of Identity: Ernest Hemingway

✳

Anders Hallengren

The recognition of Hemingway as a major and representative writer of the United States of America, was a slow but explosive process. His emergence in the western canon was an even more adventurous voyage. His works were burnt in the bonfire in Berlin on May 10, 1933 as being a monument of modern decadence. That was a major proof of the writer's significance and a step toward world fame.

To read Hemingway has always produced strong reactions. When his parents received the first copies of their son's book *In Our Time* (1924), they read it with horror. Furious, his father sent the volumes back to the publisher, as he could not tolerate such filth in the house. Hemingway's apparently coarse, crude, vulgar and unsentimental style and manners appeared equally shocking to many people outside his family. On the other hand, this style was precisely the reason why a great many other people liked his work. A myth, exaggerating those features, was to be born.

Hemingway in Our Time

After he had committed suicide at Ketchum, Idaho, in 1961, the literary position of the 1954 Nobel Laureate changed significantly and has, in a way, even become stronger. This is partly due to several posthumous works and collections that show the author's versatility — *A Moveable Feast* (1964), *By-Line* (1967), *88 Poems* (1979), and *Selected Letters* (1981). It is also the result of painstaking and successful Hemingway research, in which The Hemingway Society (USA) has played an important role since 1980.

Another result of this enduring interest is that many new aspects of Hemingway's life and works that were previously obscured by his public image have now emerged into the light. On the other hand, posthumously published novels, such as *Islands in the Stream* (1970) and *The Garden of Eden* (1986), have disappointed many of the old Hemingway readers. However, rather than bearing witness to declining literary power, (which, considering the author's declining health would, indeed, be a rather trivial observation even if it were true) the late works confront us with a reappraisal and reconsideration of basic values. They also display an unbiased seeking and experimentation, as if the author was losing both his direction and his footing, or was becoming unrestrained in a new way. Just as modern Hemingway scholarship has added immensely to the depth of our understanding of Hemingway — making him more and more difficult to define! — these works reveal and stress a complexity that may cause bewilderment or relief, depending on what perspective one adopts.

The "Hard-boiled" Style

The slang word "hard-boiled", used to describe characters and works of art, was a product of twentieth-century warfare. To be "hard-boiled" meant to be unfeeling, callous, coldhearted, cynical, rough, obdurate, unemotional, without sentiment. Later to become a literary term, the word originated in American Army World War I training camps, and has been in common, colloquial usage since about 1930.

Contemporary literary criticism regarded Ernest Hemingway's works as marked by his use of this style, which was typical of the era. Indeed, in many respects they were regarded as the embodiment and symbol of hard-boiled literature. However, neither Hemingway the man nor Hemingway the writer should be labeled "hard-boiled" — his style is the only aspect that deserves this epithet, and even that is ambiguous. Let us get down to basics, concentrate on one main feature in his literary style, and then turn to the alleged hard-boiled mind behind it, and his macho style of living and speaking.

An unmatched introduction to Hemingway's particular skill as a writer is the beginning of *A Farewell to Arms*, certainly one of the most pregnant opening paragraphs in the history of the modern American novel. In that passage the power of concentration reaches a peak, forming a vivid and charged sequence, as if it were a 10-second video summary. It is packed with events and excitement, yet significantly frosty, as if unresponsive and numb, like a silent flashback dream sequence in which bygone images return, pass in review and fade away, leaving emptiness and quietude behind them. The lapidary writing approaches the highest style of poetry, vibrant with meaning and emotion, while the pace is maintained by the exclusion of any descriptive redundancy, of obtrusive punctuation, and of superfluous or narrowing emotive signs:

> *In the late summer of that year we lived in a house in a village that looked across the river and the plain to the mountains. In the bed of the river there were pebbles and boulders, dry and white in the sun, and the water was clear and swiftly moving and blue in the channels. Troops went by the house and down the road and the dust they raised powdered the leaves of the trees. The trunks of the trees too were dusty and the leaves fell early that year and we saw the troops marching along the road and the dust rising and leaves, stirred by the breeze, falling and the soldiers marching and afterwards the road bare and white except for the leaves.*

At the end of the sixteenth chapter of *Death in the Afternoon* the author approaches a definition of the "hard-boiled" style:

> *"If a writer of prose knows enough about what he is writing about he may omit things that he knows and the*

reader, if the writer is writing truly enough, will have a
feeling of those things."

Ezra Pound taught him "to distrust adjectives" (*A Moveable Feast*). That meant creating a style in accordance with the esthetics and ethics of raising the emotional temperature towards the level of universal truth by shutting the door on sentiment, on the subjective.

"Boiling it down always, rather than spreading it out thin."
(Letter to Mary Downey Pfeiffer, Madrid, Oct. 17, 1933.)

The Unwritten Code

Later biographic research revealed, behind the macho façade of boxing, bullfighting, big-game hunting and deep-sea fishing he built up, a sensitive and vulnerable mind that was full of contradictions.

In Hemingway, sentimentality, sympathy, and empathy are turned inwards, not restrained, but vibrant below and beyond the level of fact and fable. The reader feels their presence although they are not visible in the actual words. That is because of Hemingway's awareness of the relation between the truth of facts and events and his conviction that they produce corresponding emotions.

"Find what gave you the emotion; what the action was
that gave you the excitement. Then write it down making it
clear so the reader will see it too and have the same ⬦
feeling as you had." (By-Line)

That was the essence of his style, to focus on facts. Hemingway aimed at "the real thing, the sequence of motion and fact which made the emotion and which would be as valid in a year or in ten years or, with luck and if you stated it purely enough, always" (*Death in the Afternoon*). In Hemingway, we see a reaction against Romantic turgidity and vagueness: back to basics, to the essentials. Thus his new realism in a new key resembles the old Puritan simplicity and discipline; both of them refrained from exhibiting the sentimental, the relative.

Hemingway's sincere and stern ambition was to approach Truth, clinging to an as yet unwritten code, a higher law which he referred to as "an absolute conscience as unchanging as the standard meter in Paris" (*Green Hills of Africa*, I: 1).

Hemingway's Near-Death Experience

Though Hemingway seems to have seen himself and life in general reflected in war, he himself never became reconciled to it. His mind was in a state of civil war, fighting demons inwardly as well as outwardly. In the long run defeat is as revealing and fundamental as victory: we are all losers, defeated by death. To live is the only way to face the ordeal, and the ultimate ordeal in our lives is the opposite of life. Hence Hemingway's obsession with death. Deep sea fishing, bull-fighting, boxing, big-game hunting, war, — all are means of ritualizing the death struggle in his mind — it is very explicit in books such as *A Farewell to Arms* and *Death in the Afternoon*, which were based on his own experience.

Modern investigations into so-called Near-Death Experiences (NDE) such as those by Raymond Moody, Kenneth Ring and many others, have focused on a pattern of empirical knowledge gained on the threshold of death; a dream-like encounter with unknown border regions. There is a parallel in Hemingway's life, connected with the occasion when he was seriously wounded at midnight on July 8, 1918, at Fossalta di Piave in Italy and nearly died. He was the first American to be wounded in Italy during World War I. Here is a case of NDE in Hemingway, and I think that is of basic importance, pertinent to the understanding of all Hemingway's work. In *A Farewell to Arms*, an experience of this sort occurs to the ambulance driver Frederic Henry, Hemingway's alter ego, wounded in the leg by shellfire in Italy. (Concerning the highly autobiographical nature of *A Farewell to Arms*, see Michael S. Reynolds's documentary work *Hemingway's First War: The Making of A Farewell to Arms*, Princeton University Press, 1976). As regards the NDE, we can note the incidental expression "to go out in a blaze of light" (letter to his family, Milan October 18, 1918), and the long statement about what had occurred: Milan, July 21, 1918 (*Selected Letters*, ed. Carlos Baker, 1981).

Hemingway touched on that crucial experience in his life — what he had felt and thought — in the short story "Now I Lay Me" (1927):

> *"my soul would go out of my body ... I had been blown up*
> *at night and felt it go out of me and go off and then come*
> *back"*

— and again, briefly, in *In Our Time* in the lines on the death of Maera. It reappears, in another setting and form, in the image of immortality in the African story *The Snows of Kilimanjaro*, where the dying Harry knows he is going to the peak called "Ngàje Ngài", which means, as explained in Hemingway's introductory note, "the House of God".

The Coyote and the Leopard

Hemingway's seeming insensitive detachment is only superficial, a compulsive avoidance of the emotional, but not of the emotionally tinged or charged. The pattern of his rigid, dispassionate compressed style of writing and way of life gives a picture of a touching Jeremiad of human tragedy. Hemingway's probe touches nerves, and they hurt. But through the web of failure and disillusion there emerges a picture of human greatness, of confidence even.

Hemingway was not the Nihilist he has often been called. As he belonged to the Protestant nay-saying tradition of American dissent, the spirit of the American Revolution, he denied the denial and acceded to the basic truth which he found in the human soul: the will to live, the will to persevere, to endure, to defy. The all-pervading sense of loss is, indirectly, affirmative. Hemingway's style is a compulsive suppression of unbearable and inexpressible feelings in the chaotic world of his times, where courage and independence offered a code of survival. Sentiments are suppressed to the boil.

The frontier mentality had become universal — the individual is on his own, like a Pilgrim walking into the unknown with neither shelter nor guidance, thrown upon his own resources, his strength and his judgment. Hemingway's style is the style of understatement since his hero is a hero of action, which is the human condition.

There is an illuminating text in William James (1842–1910) which is both significant and reminiscent, bridging the gap between Puritan moralism, its educational parables and exempla, and lost-generation turbulent heroism. In a letter written in Yosemite Valley to his son, Alexander, William James wrote:

> *"I saw a moving sight the other morning before breakfast*
> *in a little hotel where I slept in the dusty fields. The young*
> *man of the house had shot a little wolf called coyote in the*
> *early morning. The heroic little animal lay on the ground,*
> *with his big furry ears, and his clean white teeth, and his*
> *jolly cheerful little body, but his brave little life was gone. It*
> *made me think how brave all these living things are. Here*
> *little coyote was, without any clothes or house or books*
> *or anything, with nothing but his own naked self to pay*
> *his way with, and risking his life so cheerfully — and losing*
> *it — just to see if he could pick up a meal near the hotel. He*
> *was doing his coyote-business like a hero, and you must*
> *do your boy-business, and I my man-business bravely,*
> *too, or else we won't be worth as much as a little coyote."*
> (*The Letters of William James*, ed. Henry James,
> Little, Brown and Co.: Boston, 1926)

The courageous coyote thus serves as a moral example, illustrating a philosophy of life which says that it is worth jeopardizing life itself to be true to one's own nature. That is precisely the point of the frozen leopard close to the western summit of Kilimanjaro in Hemingway's famous short story. That is the explanation of what the leopard was seeking at that altitude, and the answer was given time and again in the works of Ernest Hemingway.

But what about the ugliness, then? What about all the evil, the crude, the rude, the rough, the vulgar aspects of his work, even the horror, which dismayed people? How could all that be compatible with moral standards? He justified the inclusion of such aspects in a letter to his "Dear Dad" in 1925:

"The reason I have not sent you any of my work is because you or Mother sent back the In Our Time books. That looked as though you did not want to see any. You see I am trying in all my stories to get the feeling of the actual life across — not to just depict life — or criticize it — but to actually make it alive. So that when you have read something by me you actually experience the thing. You can't do this without putting in the bad and the ugly as well as what is beautiful. Because if it is all beautiful you can't believe in it. Things aren't that way. It is only by showing both sides — 3 dimensions and if possible 4 that you can write the way I want to.

So when you see anything of mine that you don't like remember that I'm sincere in doing it and that I'm working toward something. If I write an ugly story that might be hateful to you or to Mother the next one might be one that you would like exceedingly."

Merging Gender in Eden

Like many other of his works, *True at First Light* was a blend of autobiography and fiction in which the author identified with the first person narrator. The author, who never kept a journal or wrote an autobiography in his life, draws on experience for his realism, slightly transforming events in his life. In this sense, the posthumous novel *Islands in the Stream* is in some places neither fictional nor fictitious. *The Garden of Eden*, however, a book brimming with the author's vulnerability just as *A Farewell to Arms* is, treats intimate and delicate matters. It is a story told in the third person, as are all his major works. Thus we get to know the writer David Bourne, assuredly an explorer like Daniel Boone, on his adventurous Mediterranean honeymoon.

The anti-hero's wife in *The Garden of Eden*, Catherine Bourne, is one of the most persuasive and lively heroines in Hemingway's works. She is depicted with fascination and fear, like Marcel Proust's Albertine and, at least in name, she reminds us of the strong and attractive

Catherine Barkley (alias the seven-year-older Agnes Von Kurowsky), the Red Cross heroine in *A Farewell to Arms*. The former character is much more complex and difficult to define, however, and her ardor and the fire of marital love prove consuming and transmogrifying.

Living at the Grau ("canal") du Roi, on the shores of the stream that runs from Aigues-Mortes straight down to the sea, the newly wedded couple in *The Garden of Eden* live in a borderland where "water" and "death" are key words, and where connotations like *L'eau du Léthe* present themselves: Eros and Thanatos, love and death, paradise and trespass. In this innocent borderland, moral limits are immediately extended, and conventional roles are reversed. Sipping his post-coital *fine à l'eau* in the afternoon, David Bourne feels relieved of all the problems he had before his marriage, and has no thought of "writing nor anything but being with this girl," who absorbs him and assumes command. Then the blond, sun-tanned Catherine appears with her hair "cropped as short as a boy's," declaring:

> *"now I am a boy ... You see why it's dangerous, don't*
> *you? ... Why do we have to go by everyone else's rules?*
> *We're us ... Please understand and love me ... I am Peter*
> *... You're my beautiful lovely Catherine."*

From that moment the tables are turned. David-Catherine accepts and submits, and Catherine-Peter takes over the man's role. She mounts him in bed at night, and penetrates him in conjugal bliss:

> *"He had shut his eyes and he could feel the long light*
> *weight of her on him ... and then lay back in the dark*
> *and did not think at all and only felt the weight and the*
> *strangeness inside and she said: 'Now you can't tell who*
> *is who can you?"*

The Father in the Garden

Women with a gamin hairstyle, lovers who cut and dye their hair and change sexual roles, are themes that, with variations, occur in his novels from *A Farewell to Arms*, *For Whom the Bell Tolls*, to the posthumous

Islands in the Stream. They culminate in *The Garden of Eden.* When writing *The Garden of Eden* he appeared as a redhead one day in May 1947. When asked about it, he said he had dyed his hair by mistake. In that novel, the search for complete unity between the lovers is carried to extremes. It may seem that the halves of the primordial Androgyne of the Platonic myth (once cut in two by Zeus and ever since longing to become a complete being again) are uniting here. Set in a fictional Paradise, a Biblical "Eden", the novel is perhaps even more a story about expulsion, the loss of innocence, and the ensuing liberation, about knowledge acquired through the Fall, which is the basis of culture, about the ordeals and the high price an author must pay to become a writer worthy of his salt. Against a mythical background, the voice of Hemingway's father is heard, challenging his son, as did the Father in the Biblical Garden. Slightly disguised, Hemingway's dear father, who haunted his son's life and work even after he had shot himself in 1928, remained an internalized critic until Ernest also took his life in 1961. Hemingway's père pressed his ambivalent son to surpass himself and produce a distinct and lively multidimensional text, — "3 dimensions and if possible 4":

> *"He found he knew much more about his father than*
> *when he had first written this story and he knew he*
> *could measure his progress by the small things which*
> *made his father more tactile and to have more dimensions*
> *than he had in the story before."*

After they had committed honeymoon adultery with the girl both spouses equally love passionately, David exclaims: "We've been burned out ... Crazy woman burned out the Bournes." This consuming and transforming fire of love and its subsequent trials and transgressions, in the end has a purging effect on the writer, who finally, as if emerging from a chrysalis stage, rises like the Phoenix from his bed and sits down in a regenerated mood to write in a perfect style:

> *"He got out his pencils and a new cahier, sharpened five*
> *pencils and began to write the story of his father and the*
> *raid in the year of the Maji–Maji rebellion ... David wrote*

steadily and well and the sentences that he had made before came to him complete and entire and he put them down, corrected them, and cut them as if he were going over proof. Not a sentence was missing ... He wrote on a while longer now and there was no sign that any of it would ever cease returning to him intact."

Maji–Maji and Mau Mau

But why is Maji–Maji so important to the author when he has attained perfection?

When Tanzania gained independence in 1961–62, President Julius Nyerere proclaimed that the new republic was the fulfillment of the Maji–Maji dream. The Maji–Maji Rebellion had been a farmers' revolt against colonial rule in German East Africa in 1905–1907. It began in the hill country southwest of Dar es-Salaam and spread rapidly until the insurrection was finally crushed after some 70,000 Africans had been killed. The farmers challenged the German militia fearlessly, crying "Maji! Maji!" when they attacked, believing themselves to be protected from bullets and death by "magic water". Maji is Swahili for "water" — one of the key words in Hemingway's novel.

The conviction and purposefulness of the Maji–Maji in *The Garden of Eden*, corresponds to the Kenyan Mau-Mau context of the novel *True at First Light*, which Hemingway started writing after his East African safari in 1953. Mau Mau was an insurrection of Kikuyo farm laborers in 1952. It was led by Jomo Kenyatta, who was subsequently held in prison until he became the premier of Kenya in 1963 (and the first President of the Republic in 1964). For Kikuyo men or women (and there were several women in the movement), to join Mau Mau meant dedicating their lives to a cause and sacrificing everything else, it meant taking a sacred oath that definitely cut them off from decorum and ordinary life.

In Hemingway's vision, Maji–Maji and Mau Mau blend with his notion of the ideal committed writer, a man who is prepared to die for his art, and for art's sake.

In the private library of Dag Hammarskjöld, who was awarded the Nobel Peace Prize after his death in the Congo (Africa) in 1961, the year Hemingway died, a copy of the beautiful original edition of *A Farewell to Arms* (Charles Scribner's Sons, 1929) may still be seen (now in the Royal Library, Stockholm). In a way it is significant that the Secretary-General of the United Nations, who was dedicated to peacemaking, should have been a Hemingway reader.

✳

"Close to the western summit
there is the dried and frozen carcass
of a leopard. No one has explained
what the leopard was seeking at
that altitude."

The Snows of Kilimanjaro

Illustration by A. Andersson.
© *The Nobel Foundation.*

Grazia DELEDDA (MADESANI, Grazia, nee Deledda)
Photo courtesy of the Nobel Foundation.

Voice of Sardegna — Grazia Deledda

✳

Anders Hallengren

Alighting in a Nordic Winter Night in 1927

At 6.45 p.m., during the lunar eclipse of an exceptionally dark and frosty winter evening on December 8, 1927, a small, Italian woman arrived at Stockholm Central after a three-day trip by train and ferry. This was her first trip to northern Europe. From her passport, issued in Rome just two weeks before her arrival in Stockholm, we get a generic picture of the shy woman who would soon be the center of attention: Height: 1.55 m; Age: 56; Eyes: Chestnut; Hair: White; Complexion: Rosy; Date of Birth: September 27, 1871. Place of Birth: Nuoro, Sardinia.

GRAZIA MARIA COSIMA DAMIANA DELEDDA, married MADESANI, was jubilantly received at the station by a committee headed by the poet Erik Axel Karlfeldt (Nobel Laureate of 1931), the permanent secretary of the Swedish Academy. Deledda was greeted with bouquets in the national colors of Sweden and Italy.

As sometimes happens, the Nobel announcement of the prize for literature surprised some people, but this amazingly popular Italian author had been among the nominees for a number of years. Her first nomination recorded by the Academy dates from 1913, when she was put forward by Italian academics. Indeed, her name had been mentioned continuously for almost two decades. Deledda had devoted readers in the Swedish Academy and among literary critics, many of whom knew Italian. She was suggested as a candidate by one of her Swedish translators, Karl August Hagberg, and repeatedly by the Swedish minister in Rome, Carl Bildt. In the early twentieth century, Deledda's international reputation as a writer had been solidly established by novels such as *Elias Portolu* and *Cenere* ("Ashes"), both published in book form in 1903.

The Private and the Public

At the time of the announcement, Deledda lived a quiet life in Rome, caring for her adult sons, the eldest named Sardus after the legendary founder of Sardinia, and her niece Grazia. When informed that she had won the Nobel Prize, the assuming woman said simply, *Già!* (Already!) and proceeded to her office to continue her regular writing schedule. She had just finished a new novel, *Annalena Bilsini* (1927), which she was adapting for the scene, and was already well into her next, *Il vecchio e i fanciulli* (The Old and the Young), which was to be published in 1928.

At home, Deledda had a pet crow called Checcha (onomatopoetic, to be sure, but also suggestive of *checché,* "whatever" and reminding of the evasive Cheshire cat in *Alice's Adventures in Wonderland*). When journalists and photographers crowded the house the following day, they were astonished to find Checcha fluttering through the rooms. When the crow finally escaped the hullabaloo and flew away, Deledda asked the visitors to leave also, so that the bird would return. "If Checcha has had enough, so have I," she reportedly said as she showed her guests to the door.

In Stockholm, the scene was not so tranquil. She was fascinated by Sweden, as is evident from her letters home and the report of her Stockholm trip that she published in *Corriere della Sera*. But she was also amazed at being surrounded by dignitaries, royalty, ambassadors, and ministers of state and felt almost dwarfed by everyone she met. It all seemed to her to be a scenario out of an old fairy tale, however, so she did not lose her head amid all the pomp and celebration. At the Nobel ceremony, when the literary historian Henrik Schück solemnly praised her in a long, incomprehensible speech delivered in her honor and she breathlessly heard her name announced and knew she was supposed to rise and approach the king to receive the prize — at that moment she thought she heard Checcha cawing. Writing to her son Franz a day after, she reminded him to feed the bird and take good care of it.

Deledda left Stockholm on December 15. Back in Italy, publicity and public attention were harder to face, as the fairy tale turned into hard reality. Benito Mussolini, who had recently come to power, wanted to profit from Deledda's fame. The writer felt compelled to participate in an embarrassing ceremony where the mark of honor awarded to her was a portrait of Mussolini with a dedication that Il Duce proudly read aloud to the audience: "For Grazia Deledda with profound admiration from Benito Mussolini."

In private, Grazia Deledda referred to these Fascist festivities as a farce, alien to her nature; but they appeared to her to be inescapable, the price of fame. Once, Mussolini asked her if he could do anything for her; she immediately requested the release of her friend and fellow-countryman Elia Sanna Mannironi, imprisoned for anti-Fascist activities. She also seems to have shared sympathies with another jailed Sardinian, ANTONIO GRAMSCI, known for his *Prison Notebooks*. But the worst embarrassment to her — and to the Swedish Academy as well — was the rumors abroad that suggested her prize was an act of political ingratiation, arranged by diplomats, this despite the fact that her name had been in nomination even before the Fascist Party was founded in 1915, as has been noted.

A Life in Books

Grazia Deledda was to live for another ten years after receiving the Nobel Prize, years marked by a painful and slowly spreading breast cancer — the incurable malady of her protagonist Maria Concezione in the fine novel *La chiesa della solitudine* (The Church of Solitude). The novel was her last, published in the year of her death. Deledda died on August 15, 1936.

Despite her disease, Deledda kept to her schedule, beginning the day with a late breakfast, hours of reading, rest after lunch, and then writing for two or three hours in the afternoon, seven days a week, year after year. She produced four handwritten pages each day. Her writing was her life. She was a quiet and reserved woman, who did not speak much. She enjoyed friendly, intimate talk and traditional feasts and celebrations, but not political debates, serious discussions, parties, or society. Yet, in her quiet way, she was gathering the material of her books, listening and observing intently, just as she had done since her childhood. The outcome was over thirty novels and some four hundred short stories, most of them collected in nineteen books. She also wrote many articles, a few plays, an opera libretto, and poems.

Even after her death, she seemed to continue to produce books. In a drawer, there was found the carefully stored manuscript of the novel *Cosima*, written in ink on light-blue paper. The book was published posthumously. Its eponymous heroine was named after the author herself, whose middle name was Cosima, and the autobiographical tale tells of Deledda's life until her first trip by train, to the capital Cagliari in southern Sardinia on October 21, 1899. That journey resulted in her marriage in January 1900 to Palmiro Madesani, a state official, and a new life in Rome. *Cosima* recalls the first half of Deledda's life, the Sardinian world that is the soul of her writings. It also explains how she became an author.

Sardinia, a Land Apart

Sardinian villages have been isolated from one another through the centuries. This is particularly true of the town of Nuoro, standing on high ground at the foot of Monte Ortobene, and the surrounding Barbargia area in the mountainous and once thickly wooded center of the island, with a rich and peculiar fauna. Sardinia (Sardegna) has a language of its own, Sardo, with many dialects. And, within the Sardinian dialects, the Nuoro dialect is special.

Thus, Grazia Deledda's mother tongue was not standard Italian but *logudorese sardo*, an Italian dialect that can be regarded as another Roman language. Deledda grew up with Sardinian legends and folklore and native customs that preserved cultural traits and themes from ancient times. For cultural even more than historical reasons, Deledda called her dear Nuoro "a bronze-age village." Its geographic location also explains another peculiar fact: the island girl Grazia Deledda never saw the sea during her childhood years.

Deledda was born on the first anniversary of the unification of Italy, and going to school and learning to read and write thus meant learning a "foreign" language, the language of a distant Italy, a language much different from the spoken idiom of her native Sardinia. Yet, despite her limited schooling in this tongue, it was to be the language in which she produced all of her written works. She became, as it were, a writer in a foreign language, *una paràula furistera,* as it is said in Logudorese — or, in Italian, *parola straniera.*

Deledda's achievement is even more remarkable because her official education lasted only four years and was on the level of primary school. This education was considered appropriate for girls at that time. The odds against such a writer's becoming a Nobel laureate are high. But, again, the posthumous autobiographical novel *Cosima* provides another explanation: the writer's very nature. For Cosima, Deledda observed, poems and short stories were written *come constretavi da una forza sotteranea* (as if forced by an unearthly power).

Childhood Gifts

Indeed, Deledda's (and Cosima's) childhood circumstances were fortunate and favorable for the development of her genius. Her family home in Nuoro, facing the majestic mountain and overlooking a vast valley, was on the way of many travelers, who often stayed over at the house. Her home, dominated by a large kitchen with a smoking *focolare* in the middle, was a center of storytelling and various encounters with fate, crime, tragedy, and romance. A well-behaved and quiet girl among more troublesome brothers and sisters, Grazia was often ignored by her busy mother Francesca and left to her own devices. The young girl kept close to her father and took pleasure in listening to and observing his many guests. Her dear, blue-eyed *Babbo* (father), the landowner and miller Giovanni Antonio Deledda, was himself a book lover and a poet, who once founded a printing office to publish a small newspaper and his own poems.

Her maternal uncle, the canon Sebastiano Cambosu, was a learned clergyman who could converse in Latin with foreigners. Cambosu observed the bright child's fascination with the magic of words and tutored her, teaching her to read and write a little before she began first grade.

When her formal schooling began, Deledda's favorite place on the way to the old convent school was Signor Carlino's libreria at Bia Maiore (today Corso Garibaldi), with its *cose magiche* (magic things) — notebooks, pens, and ink — which could *tradurre in segni la parola* (translate the word into signs); and even more than an individual word, these magic instruments could set down whole thoughts, ideas, and stories. At school, where she was permitted to jump over the first grade and was to remain the head girl in her class, Deledda was bewitched by the blackboard with its white signs, which appeared to her as a window opening out onto the image of a starlit night.

Deledda's family provided a love of nature as well as a love of literature and storytelling. Her maternal grandfather ANDREA CAMBOSU lived in his late years like a hermit, conversing with nature and surrounded by faithful animals. His children — her own mother and her teacher Don Sebastiano — were distracted dreamers, as Deledda observed in retrospect; and when they spoke, they used words "with the cutting edge of truth." There was a sort of Franciscan piety in the environment, present in naming customs as well as in the church. After a dream about her grandfather, Grazia noted that his life was full of Franciscan virtue, a warm affection that was reciprocated by the animals.

Deledda thought that the best Christmas gift she ever got was a moufflon, the shorthaired grayish-brown or russet wild sheep native to Sardinia and Corsica. It was brought to her by her *compare*, godfather, FRANCESCO SATTA from Olzai. At her birth, Satta was said to have cast her horoscope before the parents, promising an artistic or literary career. A peculiar future for a girl, and somewhat unexpected among farmers, herdsmen, and hunters in the Barbagian wilderness. This might be an important piece of information, whether exactly true or not. Daydreams of that kind may have been common among these rural people, since the family names of the three persons present at that moment in the fall of 1871 — DELEDDA (father), CAMBOSU (mother), SATTA (godfather) — were all to be prominent names in the history of Sardinian literature (Paola Pittalis, *Storia della Letteratura en Sardegna*, Cagliari: Edizione Della Torre, 1998). Their descendants — like the famous authors SEBASTIANO SATTA (1867–1914) and SALVATORE CAMBOSU (1895–1962), as well as the Nobel Laureate herself — realized their dreams. That may be the truth of prophecy.

Early Joys, Early Sorrows

But like the lives of most peasants, Grazia's life also had its hardships. In her childhood, bandits flourished in the Barbagia region, homeless outlaws and vagabonds. The stories of their adventures, crimes, and misfortunes filled the minds of the children with excitement and inspired courage. Cosima "felt the instinct of Amazons," Grazia wrote of her alter ego. But there was also terror.

When she received the moufflon, she was nine. It was the hardest winter in living memory. Francesco Satta appeared with the moufflon in his arms. He had been robbed by bandits, who had taken his horse and his winter clothes, the *mastrucca*, his fleece vest. But he was elated. He was frostbitten, but he lived. And he had got back the wild moufflon, which had run away when he was attacked.

That winter the snow buried mountains and towns. In one night, it rose more than a meter around the Deledda house. Many people in the neighborhood starved or froze to death in the terrible spell of cold, and the spell of human suffering Grazia took to her heart that winter never left her. Families flocked together by the fireplaces in the kitchens; death took its toll in the Deledda home too. Her little sister Giovanna, three years younger than herself, was found dead one day in her bed. Grazia always remembered her as the most beautiful of her sisters. That winter the church bells often chimed in the winds. There is no reason to doubt that Grazia's dear Christmas gift was consumed out of necessity. That winter changed everything.

Grazia also observed the complex rituals of marriage in the bright season, the days of *paralimpos* processions carrying gifts to the new home, including the mattress that would witness the birth of the couple's children and their own deaths, and the embroidered *inghirialetto*, the sacred bedspread of fertility and happiness. And she observed the food offerings placed at the doorways of many houses at Christmas, sacrifices signaling the loss of dear ones. Love and death were intertwined in the human fabric.

Her paternal grandfather, Santus Deledda was called Su Santain, "the saint-maker," because he was famous for his wooden statues of saints, displayed at the religious festivals in Nuoro. In this Catholic context, the ancient gods of Sardinia were still present. Among them was Molk, who demanded burnt offerings and who came to people's minds when forest fires raged at Monte Ortobene. Death and sacrifice thus in a complex way became symbolized by ashes, *cenere*.

The mountain was enchanted. The most exciting moments of Grazia's childhood days were the horseback rides through the *tancas* (enclosures) of valleys and ridges with her older brother Andrea (born in 1866). He brought her uphill to *Domus de janas*, the Tomb of the Giant, the strange rock coffin covered with moss, solemn in the vast solitude of a place alive with ancient legends. She identified with the fairies, the little women of the mountain caves, who for thousands of years had been weaving nets on their golden looms to imprison hawks, winds, clouds, and dreams. And she heard the trees speak in the wind, murmuring "Why?" Wind songs in the distance transported her vague sadness on the wings of their choir.

The heights were also the site of joyous occasions, like the *Novena* at the little cliff church called Madonna Del Monte, where her uncle Ignazio served as the parish priest. Festivals like the Feast of the Redeemer went on for several days, with the congregation spending nights in small houses around the church. It was a meeting place for the young. Since the middle ages, churches have served as social centers and have been the starting point of romances: the place where young men and women from the villages saw each other and where erotic attractions were born. To the teenager, nature turned into an enchanted poem of love and adventure.

The summit and turning point of her teens was her climbing Monte Bardia on a long ride eastward from Oliena to Dorgali. The mountains and the stories of its shepherds had delighted her from her earliest days. Finally, one day on her own she climbed the peak and saw the Mediterranean for the first time. She felt humble but experienced an awakening. The sea came to her as a singular revelation and opened up new vistas to her life. From a boulder overlooking the sea, the Mediterranean looked like a shining sword that had sliced her island from the continent in a distant past. A dream of Rome and of becoming a writer rose in her mind from the depths of her soul. In her own words, she then plunged headlong "into a sea of visions."

Teenage Love Stories

In 1887, the young writer completed her first short story, "Sangue Sardo" (Sardinian blood) and secretly mailed it to a fashion magazine in Rome, the *Ultima Moda*, which published short pieces of fiction. "Sangue Sardo" was a story about a girl like Grazia involved in a love triangle and its jealousies. Set by the sea, the story ends in murder when the protagonist Ela pushes her sister's lover from the cliffs. Being a published writer of love stories at seventeen, Deledda found more infamy than fame in her village. Suspicion and rumors followed her. Her mother was attacked for being an irresponsible parent; village women burned a magazine and shouted their reproaches. To deflect the shock and anger engendered by her fiction, Deledda began publishing under pseudonyms such as *G. Razia* or even the biblical *Ilia di Sant'Ismael* when she was published in local magazines. But the stories now followed with an irresistible force: "Remigia Helder" (1888), the collection *Nell'azzurro* (1990), and her first novels: *Memorie di Fernanda* (1888), *Stella d'Oriente* (1890), *Amore regale* (1891), *Amori fatali* (1892), *Fior di Sardegna* (1892).

With *Fior di Sardegna* (The Flower of Sardinia), Deledda became famous; but in Nuoro, not even Signor Carlino's book store, the magic place of her childhood, accepted her volumes. She identified with her protagonists who were created from real life, from people she either knew or had heard of. Local people, however, continued to identify these secret lovers on nightly errands — like the yearning Lara in *Fior di Sardegna* — with the author. And there was some truth to this. As she later revealed as the secret of her writings, all the agonies of her characters were her own suffering, her own pain, and her own tears. And these streaks of blackness were constantly widening.

As became more and more evident, Deledda's aim in her art was to picture the life, the sentiments, and the thoughts of her culture on a broader scale, and to set in writing the stories of her island. In the following years, when she was in her early twenties, she collected folklore as a scholar, partly in collaboration with ANGELO DE GUBERNATIS and the *Rivista delle Tradizioni Popolari Italiane* in a research project that resulted in publications such as *Tradizioni popolari di Nuoro in Sardegna* (1895). Fictional outcomes of this period of studies were the anecdotal and ironic *Anime oneste*, "Honest Souls," (1895), and the serious and socially penetrating *La via del male* (1896), "the way of evil."

From this point on, all Deledda was to write developed into an inquiry into moral conflicts, human suffering, and the problem of will and destiny. Identification turned out to be an act of empathy.

Family Tragedies

Indeed, Deledda did not have to look far for her models of love and tragedy; her own family provided ample material for her themes of human suffering. Her sister Enza (=Vincenza, b. 1868) had a secret love, a disgrace to the family until a wedding was finally arranged. But one day, Enza was found dead in her bed, lying in a pool of blood. She had died during a miscarriage. Grazia took care of her sister's body, closing her eyes, cleaning the bed, and washing her limbs, shrouding her carefully as a bride in white. She perfumed Enza and arranged her beautiful chestnut hair on the wedding bed. After that day spent in solitude with her sister's body, sorrow followed Deledda like an inescapable terror.

Her brother Andrea, once a promising student, proved to be the family black sheep. He stole money from his father to visit prostitutes in San Pietro, the poorest part of the little town. He fathered an illegitimate child with a neighbor girl, and he committed new thefts. He swore to hang himself in prison if he was arrested one more time, but he was imprisoned again and did not keep his word. When their *Babbo* died in 1892, Andrea wasted the family money. So, even into their own house "came deceitful, poisonous and perhaps unavoidable evil," as happens also in *Cosima*.

Another brother, Santus (b. 1864), was the opposite of Andrea, but he too would come to a bad end. Santus did not drink or chase girls, but was studious and inventive. One of his triumphs was a hot-air balloon of silk paper with a load of burning charcoal. It rose from the garden and flew above the town, drawing much attention before it disappeared among the mountains. Several days later, the family learned that it had come down on a mountain ledge after nightfall; terrified goat herders fell down on their knees in prayers, believing it was the Holy Spirit descending.

Santus' experiments with fireworks were partly successful too, until the inventor was severely burned in an accident. Suffering from the pain of his extended injuries and disheartened by his various failures, Santus started drinking. Grazia remembered him as fine and vulnerable, but "he cracked like a crystal cup or porcelain vase cracks by a blow." Finally, after his last attempts to continue his studies in Cagliari, he one night knocked at the door of the family home in Nuoro "like a dead man going down the street knocking on doors to warn the living that hell is near."

Santus' delirium led his mother to believe that he was possessed by an evil spirit. Soon she herself sank deeper and deeper into melancholy and depression. Grazia took over the responsibility of the Deledda business, the local olive oil press and the bookkeeping.

Life Is a River

Faced by the problems of her brothers and the unhappiness of her whole family, Deledda became fatalistic. She more and more saw life as a river, changing character as it passed on its way, from the upper reaches to the lower. *La vita segue il suo corso fluviale*, "Life follows the course of its flow": calm periods alternate with turbid ones. In vain we try to raise dams or even to lay ourselves across the current to stop its flow. We are powerless against outward and inward forces, sometimes making things even worse by fighting against them. These forces of evil and misery, haunting human existence, are incomprehensible, whether we look at them as chance, fate, or divine providence. This is the tragic mystery of being. *Forze occulte, fatali, spingono l'uomo al bene o al male; la natura stessa, che sembra perfetta, é sconvolta dalle violenze di una sorte ineluttabile*; "occult and fatal forces drive us to good or evil; even nature herself, which appears to us perfect, is violently turned upside-down by inevitable destiny."

Despite this fatalism, a main theme of Deledda's novels is moral conflicts, transgressions, and private revolts, the battle of free will against fate. This is the case with the love theme of *Amori fatali* (1892), the unforgettable novel *La via del male*, and the precarious dilemmas of *La giustizia* (1899) and *Le tentazioni* (1899). The mysterious driving forces of our souls compel us to act, so that transgression becomes an almost involuntary act. Our will may be free, but we cannot command our will; we cannot even decide what we want. Thus we are not masters of our lives but are as *canne al vento*, "reeds in the wind." Even the natural gifts of the artist are not matters of personal pride, since they are bestowed by grace and a sometimes cruel taskmaster, a lifelong mission propelled "by an unearthly power." The dilemma — and tragedy — is that, despite this buffeting of incomprehensible fate, we remain responsible for our acts.

The Lot of Elias

As if happy to escape, Deledda married an *istranzu* (stranger) and left her island for Rome. Yet the island belief *kie venit dae su mare*, that all strange things "come from across the sea," never left her. Her best novels on Sardinian life were written after her departure. During this intense period of new love and conflict on leaving her family home, Deledda wrote the nostalgic and mythic story of the goat-herd of

Monte Ortobene, *Il Vecchio della Montagna*. She also began writing *Elias Portolu*, a novel partly named after another shepherd in their neighborhood, Elias Porcu. On a superficial level, the theme of *Elias Portolu* is reminiscent of VICTOR HUGO's *Les Misérables*, a novel twice referred to in the autobiographical *Cosima* and obviously one of Deledda's favorites. But the models for the character were drawn from life, not from literature, as was the whole milieu of the plot — shepherds, local rites, and customs, the old wise man of the mountain whose advice should have been followed, the church at the rock, animals used as metaphors of human traits, and the stern presence of Catholicism. She told the story of her brother Andrea, returning from jail, the father of an illegitimate child. She had many other relatives in mind when she wrote this masterpiece of tragic love, among them her skeptical Uncle Don Ignazio, known to be a clergyman weak in faith and fond of worldly things. These complex characters are actors in a drama of strong moral force, fighting against their yearnings and their fates.

The convict Elias Portolu returns to Nuoro determined to reform his life but destined to fail. He cannot destroy his secret love for his brother's wife Maddalena, which ends in adultery and the birth of a child, which is believed to be his brother's son. Elias decides to become a priest to atone for his sin and to live a better life. He remains tormented by his love for Maddalena and their son, who are badly treated by his drunken and violent brother. But when the brother dies, Elias's conscience prevents him from joining his true family. At last, at the early death of his son, Elias receives salvation.

The novel was finished in the summer of 1900, when Grazia Deledda Madesani had settled down in Rome after the honeymoon and was pregnant. She experienced immense happiness after giving birth to her son Sardus and then to Franz (b. 1904); yet, in the midst of this wonderful period, she wrote *Dopo il divorzio*, "After the divorce," (1902) and began to produce the long progression of serious novels that form the summit of her life as a writer. These works seethe with pity and compassion for her protagonists, drawn from life: *Cenere* (1903), *Nostalgie* (1905), *L'ombra del passato* (1907), *L'edera* (1908), *Sino al confine* (1910), *Nel deserto* (1911), *Colombi e sparvieri* (1912),

Canne al vento (1913), *Le colpe altrui* (1914), *Marianna Sirca* (1915), *L'incendio nell'oliveto* (1918), *La madre* (1920) *Il segreto dell'uomo solitario* (1921).

The themes of all these works are related, and the stories, seen as a whole, intertwine and focus on moral dilemmas, passions, and human weakness. The uncontrollable forces of life and the human condition check freedom of the will, and in this tangle, the author investigates the anatomy of human tragedy. As in *Elias Portolu*, good intentions often result in bad decisions, and conscience proves to be a disastrous inhibitor. It seems that the only action worse than submitting to fate is opposing fate, when things get even worse. A war rages between nature and culture, and individuals are victims.

"To be reduced to ashes"

Cenere (Ashes), her most disturbing study of human tragedy, tells of a poor "fallen" woman, who, making a grievous moral sacrifice, leaves her illegitimate child with foster parents to give him a better chance in life. Wishing him all happiness in the world, far from her own misery and sinfulness, she gives her little child a sacred *rezetta* amulet and leaves. However, her son searches and longs for his mother all his life. The amulet, serving as a unifying symbol of the plot, identifies the son when he finally finds his old and ailing mother, abandoning his prospects and ending his engagement to a young woman, only to drive his mother to suicide with his reproaches. This novel, certainly Deledda's strongest, captivated readers throughout Europe and was adapted for the screen. *Cenere* (1916) was directed by Febo Mari and filmed on location in Sardinia. The movie has a special significance, due to ELEONORA DUSE's (1858–1924) strangely empathetic interpretation of the mother. Duse came out of retirement to make her one and only appearance in a film. In this unique classic in the history of the silent movie, Duse creates the role of the mother who left her son in order to shield him, a sacrifice the actress had in fact once made herself.

Deledda's recurrent image of corruptibility is also found in the short story "While the East Wind Blows," in which the peasants harbor the futile dream of not being reduced to ashes, *non essere ridotti in cenere*. Taking place on Christmas Eve, the story depicts the eternal triangle of love, birth, death.

"We are like reeds in the wind"

Although the futility of human striving was a constant theme in Deledda's work, there are also images of something deeply rooted that survives through the ages. The fascinating novel *Canne al Vento*, "Reeds in the Wind," is framed by the mythic and elemental beings of Sardinian fairyland and set on the Sardinian eastern coast. The novel depicts this deeply seated constancy from the beginning to the end. With its biblical reference — "But what went ye out in the wilderness to see?" Jesus asked the multitudes assembled. "A reed shaken with the wind?" (Matthew 11:7) — Deledda answers Jesus' question in the negative.

In *Canne al Vento*, the Pintor sisters live out their isolated lives decade after decade in the house in Galte (Galtelli), where old secrets torment the characters until death. They enact an archetypal drama. The wise old Efix, repenting for ancient sins, is an image of humanity and represents continuity. The undercurrent of Deledda's novels is a deep faith in humanity and in the truth of the human heart. This book, as with most of her novels, implicitly criticizes moral norms and social values, but does not criticize the people who are caught up in their web.

Once more addressing the theme of suffering mothers in her popular novel *La Madre* (1920), tragedy and sorrow are the natural outcomes of love. The moral ambiguity of the theme in *The Mother* is indicated by the name of the strong and well-meaning hero, Maria Maddalena. In a remarkable study, Maria Giovanna Piano showed the mother's variety of conflicting aspects (*Onora la madre: Autorità femminile nella narrativa di Grazia Deledda*, Turin: Rosenberg & Sellier, 1998). This is not a homage to motherhood but first and foremost an appeal to mercy, forgiveness, and understanding. The mother is a flawed human being who cannot handle the secret fact that her son is both a catholic priest and a sensual lover. Bigotry and social norms are attacked, but not the transgressors, who suffer from their instincts. Having attained her goal to immortalize the society she came from, Grazia Deledda ended up as one of its severest critics.

Influences and Rebuffs

Italian writers of Deledda's generation wrote in the vast shadow of the impressive GABRIELE D'ANNUNZIO (1863–1938), a critical giant, admired by the public and adored by the intelligentsia. D'Annunzio was idolized by Grazia's brother Andrea and his fellow-students. However, despite D'Annunzio's reputation, he was not an important influence on Deledda. Grazia obviously read him and may have been attracted by his mythical interpretation of reality and personal identity, but she never admired him. Rather, she listened to folktales and folksongs, preferring the local *stornelli* ballads of troubled women to all the writings of d'Annunzio or classics like Dante.

A more profound literary impact blew on the winds of another island, Sicily. There she found a neighboring literature that was down-to-earth, represented by naturalists such as LUIGI CAPUANA (1839–1915) and GIOVANNI VERGA (1840–1922). These writers minutely observed local reality and carefully set their characters and their actions into a social context. Grazia wanted to be a Sardinian counterpart. Capuana aimed at producing fiction that was absolutely true to reality. Verga's objective was to demonstrate human events in a somewhat scientific manner, describing common people. Verga's protagonists were victims of their circumstances, objectively described. Capuana himself was to praise Grazia Deledda for the social realism of her novel *La via del male*.

When she received the Nobel Prize thirty years afterwards, Deledda mentioned Verga as a worthier winner than herself. She was to draw much from contemporary currents of *verismo*, which curbed the romantic and intimate traits of her early stories, forever keeping a deep sense of compassion and empathy, however, quite contrary to the programmatic objectivity of naturalists and realists. Nevertheless, these writers inspired her to create a genuine Sardinian literature on her own and become the Italian voice of her island.

Other significant influences on Deledda's development were some pioneers of modern Italian literature: nestor ALESSANDRO MANZONI (1785–1873), who developed the novel into a modern epic art; OGO TARCHETTI (1839–1869) who opposed classic themes and ideals of beauty and focused on decay and horror in his environments; and ANTONIO FOGAZZARO (1842–1911) whose deep concern was psychology, dreams and mysteries, culminating in the passionate *Il misterio del poeta* (1888).

And then there is a Sicilian of an even more problematic identity-seeking nature: Luigi Pirandello (Nobel Prize 1934), who — perhaps unintentionally — delivered a killing blow against his Italian compatriot Deledda. The author of the dramatic masterpiece *Sei personaggi in cerca d'Autore*, "Six Characters in Search of an Author," had written a satirical novel entitled *Suo Marito*, "Her Husband," (1911), which depicted the ridiculous private life of a famous woman author named Silvia Roncella and her easygoing husband-manager. It was a roman à clef, or at least written in such a way that the contemporary audience immediately identified the Roman couple as Palmiro Madesani and his wife Grazia Deledda. Indeed, she recognized herself and was shocked at intimate details, which indicated that a common acquaintance (an acquaintance or, even more insidiously, her husband) must have been a source of information. The serious and retiring woman author, depicted as indifferent and socially awkward, never quite recovered from having figured in such drollery and isolated herself in her home. This scandal haunted both Deledda and Pirandello for the rest of their lives, both dying in 1936. One of the last manuscripts Pirandello put his hand to was a thoroughly revised version of that satire, edited for posterity, giving it a more positive note and screening it with a new title: *Giustino Roncella Nato Boggiòlo* (1941).

"All roads lead to the human heart"

The shy and reclusive Deledda let her writing speak for her. She rarely spoke in public or delivered a speech. At public appearances and receptions, the honored author was taciturn. Even at the Nobel Ceremony in Stockholm, her acceptance speech, delivered at the Grand

Hotel banquet, was one of the shortest ever. She began by excusing herself, saying that she did not know how to make a speech, and then uttered a few lines of *salute!* and thanks.

In a short radio recording from her later years in Rome, the author reads a few sentences about herself — about her origins, her family, and about their literary interests. In these preserved seconds of sound, we hear her soft voice and the warmth of her humanity, the presence of the calm and humane narrator of her Sardinian stories.

> *"I was born in Sardinia. My family consisted of wise as well as violent people, and primitive artists. The family was respected and of good standing, and had a private library. But when I started writing at thirteen, they objected. As the philosopher says: If your son is writing poems, send him to the mountain paths; the next time you may punish him; but the third time, leave him alone, because then he is a poet."*
> (Translation by Anders Hallengren and Amedeo Cottino of the radio recording from the Swedish Academy Archives.)

Archbishop Nathan Söderblom's address at the Nobel banquet delighted the retiring author in a particular way, because he addressed her in Italian, and she could not forget his words: "In your literary work, all roads lead to the human heart ... You have seen the road sign." Spokeswoman and critic of her society, she had produced a lifework of sympathy and fellow-feeling, of compassion, from her heart saying the Lord's Prayer with all her countrymen: *Su Babbo Nostru ... Nois damos perdonu, a sos nemigos nostros bois sos peccados nostros, perdonade ...* Forgive us our trespasses.

When Deledda died in 1936, she was shrouded in the maroon velvet dress she had worn during the Nobel festivities in Stockholm ten years earlier. In a quiet spot at the foot of Monte Ortobene, close to her home in Nuoro, a memorial church was built, named after her book *Chiesa della Solitudine*. There, under the shadow of the trees she passed on her excursions uphill is the lonely Tomba Deleddiana. Her final resting place, as it were, is a novel.

❈

Chiesa della Solitudine, Nuoro.
Photo courtesy of Anders Hallengren.

Amartya SEN
Photo courtesy of the author.

Amartya Sen – Autobiography[*]

I was born in a University campus and seem to have lived all my life in one campus or another. My family is from Dhaka — now the capital of Bangladesh. My ancestral home in Wari in "old Dhaka" is not far from the University campus in Ramna. My father Ashutosh Sen taught chemistry at Dhaka University. I was, however, born in Santiniketan, on the campus of Rabindranath Tagore's Visva-Bharati (both a school and a college), where my maternal grandfather (Kshiti Mohan Sen) used to teach Sanskrit as well as ancient and medieval Indian culture, and where my mother (Amita Sen), like me later, had been a student. After Santiniketan, I studied at Presidency College in Calcutta and then at Trinity College in Cambridge, and I have taught at universities in both these cities, and also at Delhi University, the London School of Economics, Oxford University, and Harvard University, and on a visiting basis, at M.I.T., Stanford, Berkeley, and Cornell. I have not had any serious non-academic job.

My planned field of study varied a good deal in my younger years, and between the ages of three and seventeen, I seriously flirted, in turn,

---

[*] From *Les Prix Nobel*, 1998.

with Sanskrit, mathematics, and physics, before settling for the eccentric charms of economics. But the idea that I should be a teacher and a researcher of some sort did not vary over the years. I am used to thinking of the word "academic" as meaning "sound," rather than the more old-fashioned dictionary meaning: "unpractical," "theoretical," or "conjectural."

During three childhood years (between the ages of 3 and 6) I was in Mandalay in Burma, where my father was a visiting professor. But much of my childhood was, in fact, spent in Dhaka, and I began my formal education there, at St. Gregory's School. However, I soon moved to Santiniketan, and it was mainly in Tagore's school that my educational attitudes were formed. This was a co-educational school, with many progressive features. The emphasis was on fostering curiosity rather than competitive excellence, and any kind of interest in examination performance and grades was severely discouraged. ("She is quite a serious thinker," I remember one of my teachers telling me about a fellow student, "even though her grades are very good.") Since I was, I have to confess, a reasonably good student, I had to do my best to efface that stigma.

The curriculum of the school did not neglect India's cultural, analytical and scientific heritage, but was very involved also with the rest of the world. Indeed, it was astonishingly open to influences from all over the world, including the West, but also other non-Western cultures, such as East and South-East Asia (including China, Japan, Indonesia, Korea), West Asia, and Africa. I remember being quite struck by Rabindranath Tagore's approach to cultural diversity in the world (well reflected in our curriculum), which he had expressed in a letter to a friend: "Whatever we understand and enjoy in human products instantly becomes ours, wherever they might have their origin ... Let me feel with unalloyed gladness that all the great glories of man are mine."

Identity and Violence

I loved that breadth, and also the fact that in interpreting Indian civilization itself, its cultural diversity was much emphasized. By pointing to the extensive heterogeneity in India's cultural background and richly diverse history, Tagore argued that the "idea of India" itself militated against a culturally separatist view, "against the intense consciousness of the separateness of one's own people from others." Tagore and his school constantly resisted the narrowly communal identities of Hindus or Muslims or others, and he was, I suppose, fortunate that he died — in 1941 — just before the communal killings fomented by sectarian politics engulfed India through much of the 1940s. Some of my own disturbing memories as I was entering my teenage years in India in the mid-1940s relate to the massive identity shift that followed divisive politics. People's identities as Indians, as Asians, or as members of the human race, seemed to give way — quite suddenly — to sectarian identification with Hindu, Muslim, or Sikh communities. The broadly Indian of January was rapidly and unquestioningly transformed into the narrowly Hindu or finely Muslim of March. The carnage that followed had much to do with unreasoned herd behaviour by which people, as it were, "discovered" their new divisive and belligerent identities, and failed to take note of the diversity that makes Indian culture so powerfully mixed. The same people were suddenly different.

I had to observe, as a young child, some of that mindless violence. One afternoon in Dhaka, a man came through the gate screaming pitifully and bleeding profusely. The wounded person, who had been knifed on the back, was a Muslim daily labourer, called Kader Mia. He had come for some work in a neighbouring house — for a tiny reward — and had been knifed on the street by some communal thugs in our largely Hindu area. As he was being taken to the hospital by my father, he went on saying that his wife had told him not to go into a hostile area during the communal riots. But he had to go out in search of work and earning because his family had nothing to eat. The penalty of that economic unfreedom turned out to be death, which occurred later on in the hospital. The experience was devastating for me, and suddenly made me aware of the dangers of narrowly defined identities, and also of the divisiveness that can lie buried in communitarian politics. It also alerted me to the remarkable fact that economic unfreedom, in the form of extreme poverty, can make a person a helpless prey in the violation of other kinds of freedom: Kader Mia need not have come to a hostile area in search of income in those troubled times if his family could have managed without it.

Calcutta and Its Debates

By the time I arrived in Calcutta to study at Presidency College, I had a fairly formed attitude on cultural identity (including an understanding of its inescapable plurality as well as the need for unobstructed absorption rather than sectarian denial). I still had to confront the competing loyalties of rival political attitudes: for example, possible conflicts between substantive equity, on the one hand, and universal tolerance, on the other, which simultaneously appealed to me. On this more presently.

The educational excellence of Presidency College was captivating. My interest in economics was amply rewarded by quite outstanding teaching. I was particularly influenced by the teaching of Bhabatosh Datta and Tapas Majumdar, but there were other great teachers as well, such as Dhiresh Bhattacharya. I also had the great fortune of having wonderful classmates, particularly the remarkable Sukhamoy Chakravarty (more on him presently), but also many others, including Mrinal Datta Chaudhuri (who was also at Santiniketan, earlier) and Jati Sengupta. I was close also to several students of history, such as Barun De, Partha Gupta and Benoy Chaudhuri. (Presidency College had a great school of history as well, led by a most inspiring teacher in the form of Sushobhan Sarkar.) My intellectual horizon was radically broadened.

The student community of Presidency College was also politically most active. Though I could not develop enough enthusiasm to join any political party, the quality of sympathy and egalitarian commitment of the "left" appealed to me greatly (as it did to most of my fellow students as well, in that oddly elitist college). The kind of rudimentary thinking that had got me involved, while at Santiniketan, in running evening schools (for illiterate rural children in the neighbouring villages) seemed now to be badly in need of systematic political broadening and social enlargement.

I was at Presidency College during 1951 to 1953. The memory of the Bengal famine of 1943, in which between two and three million people had died, and which I had watched from Santiniketan, was still quite fresh in my mind. I had been struck by its thoroughly class-dependent character. (I knew of no one in my school or among my friends and relations whose family had experienced the slightest problem during the entire famine; it was not a famine that afflicted even the lower middle classes — only people much further down the economic ladder, such as landless rural labourers.) Calcutta itself, despite its immensely rich intellectual and cultural life, provided many constant reminders of the proximity of unbearable economic misery, and not even an elite college could ignore its continuous and close presence.

And yet, despite the high moral and ethical quality of social commiseration, political dedication and a deep commitment to equity, there was something rather disturbing about standard leftwing politics of that time: in particular, its scepticism of process-oriented political thinking, including democratic procedures that permit pluralism. The major institutions of democracy got no more credit than what could be portioned out to what was seen as "bourgeois democracy," on the deficiencies of which the critics were most vocal. The power of money in many democratic practices was rightly identified, but the alternatives — including the terrible abuses of non-oppositional politics — did not receive serious critical scrutiny. There was also a tendency to see political tolerance as a kind of "weakness of will" that may deflect well-meaning leaders from promoting "the social good," without let or hindrance.

Given my political conviction on the constructive role of opposition and my commitment to general tolerance and pluralism, there was a bit of a dilemma to be faced in coordinating those beliefs with the form of left-wing activism that characterized the mainstream of student politics in the-then Calcutta. What was at stake, it seemed to me, in political toleration was not just the liberal political arguments that had so clearly emerged in post-Enlightenment Europe and America, but also the traditional values of tolerance of plurality which had been championed over the centuries in many different cultures — not least in India. Indeed, as Ashoka had put it in the third century BC: "For he who does reverence to his own sect while disparaging the sects of others wholly from attachment to his own, with intent to enhance the splendour of his own sect, in reality by such conduct inflicts the severest injury on his own sect." To see political tolerance merely as a "Western liberal" inclination seemed to me to be a serious mistake.

Even though these issues were quite disturbing, they also forced me to face some foundational disputes then and there, which I might have otherwise neglected. Indeed, we were constantly debating these competing political demands. As a matter of fact, as I look back at the fields of academic work in which I have felt most involved throughout my life (and which were specifically cited by the Royal Swedish Academy of Sciences in making their award), they were already among the concerns that were agitating me most in my undergraduate days in Calcutta. These encompassed welfare economics, economic inequality and poverty, on the one hand (including the most extreme manifestation of poverty in the form of famines), and the scope and possibility of rational, tolerant and democratic social choice, on the other (including voting procedures and the protection of liberty and minority rights). My involvement with the fields of research identified in the Nobel statement had, in fact, developed much before I managed to do any formal work in these areas.

It was not long after Kenneth Arrow's path-breaking study of social choice, *Social Choice and Individual Values*, was published in New York in 1951, that my brilliant co-student Sukhamoy Chakravarty drew my attention to the book and to Arrow's stunning "impossibility theorem" (this must have been in the early months of 1952). Sukhamoy too was broadly attracted by the left, but also worried about political authoritarianism, and we discussed the implications of Arrow's demonstration that no non-dictatorial social choice mechanism may yield consistent social decisions. Did it really give any excuse for authoritarianism (of the left, or of the right)? I particularly remember one long afternoon in the College Street Coffee House, with Sukhamoy explaining his own reading of the ramifications of the formal results, sitting next to a window, with his deeply intelligent face glowing in the mild winter sun of Calcutta (a haunting memory that would invade me again and again when he died suddenly of a heart attack a few years ago).

Cambridge As a Battleground

In 1953, I moved from Calcutta to Cambridge, to study at Trinity College. Though I had already obtained a B.A. from Calcutta University (with economics major and mathematics minor), Cambridge enroled me for another B.A. (in pure economics) to be quickly done in two years (this was fair enough since I was still in my late teens when I arrived at Cambridge). The style of economics at the-then Cambridge was much less mathematical than in Calcutta. Also, it was generally less concerned with some of the foundational issues that had agitated me earlier. I had, however, some wonderful fellow students (including Samuel Brittan, Mahbub ul Haq, Rehman Sobhan, Michael Nicholson, Lal Jayawardena, Luigi Pasinetti, Pierangelo Garegnani, Charles Feinstein, among others) who were quite involved with foundational assessment of the ends and means of economics as a discipline.

However, the major debates in political economy in Cambridge were rather firmly geared to the pros and cons of Keynesian economics and the diverse contributions of Keynes's followers at Cambridge (Richard Kahn, Nicholas Kaldor, Joan Robinson, among them), on the one hand, and of "neo-classical" economists sceptical of Keynes, on the other (including, in different ways, Dennis Robertson, Harry Johnson, Peter Bauer, Michael Farrell, among others). I was lucky to have close relations with economists on both sides of the divide. The debates centred on macroeconomics dealing with economic aggregates for the economy as a whole, but later moved to capital theory, with the neo-Keynesians dead set against any use of "aggregate capital" in economic modelling (some of my fellow students, including Pasinetti and Garegnani, made substantial contributions to this debate).

Even though there were a number of fine teachers who did not get very involved in these intense fights between different schools of thought (such as Richard Stone, Brian Reddaway, Robin Matthews, Kenneth Berrill, Aubrey Silberston, Robin Marris), the political lines were, in general, very firmly — and rather bizarrely — drawn. In an obvious sense, the Keynesians were to the "left" of the neo-classicists, but this was very much in the spirit of "this far but no further". Also, there was no way in which the different economists could be nicely ordered in just one dimension. Maurice Dobb, who was an astute Marxist economist, was often thought by Keynesians and neo-Keynesians to be "quite soft" on "neo-classical" economics. He was one of the few who, to my delight, took welfare economics seriously (and indeed taught a regular course on it), just as the intensely "neo-classical" A. C. Pigou had done (while continuing to debate Keynes in macroeconomics). Not surprisingly, when the Marxist Dobb defeated Kaldor in an election to the Faculty Board, Kaldor declared it to be a victory of the perfidious neo-classical economics in disguise ("marginal utility theory has won," Kaldor told Sraffa that evening, in commenting on the electoral success of a Marxist economist!)

However, Kaldor was, in fact, much the most tolerant of the neo-Keynesians at Cambridge. If Richard Kahn was in general the most bellicose, the stern reproach that I received often for not being quite true to the new orthodoxy of neo-Keynesianism came mostly from my thesis supervisor — the totally brilliant but vigorously intolerant Joan Robinson.

In this desert of constant feuding, my own college, Trinity, was a bit of an oasis. I suppose I was lucky to be there, but it was not entirely luck, since I had chosen to apply to Trinity after noticing, in the handbook of Cambridge University, that three remarkable economists of very different political views coexisted there. The Marxist Maurice Dobb and the conservative neo-classicist Dennis Robertson did joint seminars, and Trinity also had Piero Sraffa, a model of scepticism of nearly all the standard schools of thought. I had the good fortune of working with all of them and learning greatly from each.

The peaceful — indeed warm — co-existence of Dobb, Robertson and Sraffa was quite remarkable, given the feuding in the rest of the University. Sraffa told me, later on, a nice anecdote about Dobb's joining of Trinity, on the invitation of Robertson. When asked by Robertson whether he would like to teach at Trinity, Dobb said yes enthusiastically, but he suffered later from a deep sense of guilt in not having given Robertson "the full facts." So he wrote a letter to Robertson apologizing for not having mentioned earlier that he was a member of the Communist Party, supplemented by the statement —

I think a rather "English" statement — that he would understand perfectly if in view of that Robertson were to decide that he, Dobb, was not a fit person to teach Trinity undergraduates. Robertson wrote a one-sentence reply: "Dear Dobb, so long as you give us a fortnight's notice before blowing up the Chapel, it will be all right."

So there did exist, to some extent, a nice "practice" of democratic and tolerant social choice at Trinity, my own college. But I fear I could not get anyone in Trinity, or in Cambridge, very excited in the "theory" of social choice. I had to choose quite a different subject for my research thesis, after completing my B.A. The thesis was on "the choice of techniques," which interested Joan Robinson as well as Maurice Dobb.

Philosophy and Economics

At the end of the first year of research, I was bumptious enough to think that I had some results that would make a thesis, and so I applied to go to India on a two-years leave from Cambridge, since I could not — given the regulation then in force — submit my Ph.D. thesis for a degree until I had been registered for research for three years. I was excitedly impatient in wanting to find out what was going on back at home, and when leave was granted to me, I flew off immediately to Calcutta. Cambridge University insisted on my having a "supervisor" in India, and I had the good fortune of having the great economic methodologist, A. K. Dasgupta, who was then teaching in Benares. With him I had frequent — and always enlightening — conversations on everything under the sun (occasionally on my thesis as well).

In Calcutta, I was also appointed to a chair in economics at the newly created Jadavpur University, where I was asked to set up a new department of economics. Since I was not yet even 23, this caused a predictable — and entirely understandable — storm of protest. But I enjoyed the opportunity and the challenge (even though several graffitis on the University walls displayed the "new professor" as having been just snatched from the cradle). Jadavpur was quite an exciting place intellectually (my colleagues included Paramesh Ray, Mrinal Datta Chaudhuri, Anita Banerji, Ajit Dasgupta, and others in the economics department). The University also had, among other luminaries, the immensely innovative historian, Ranajit Guha, who later initiated the "subaltern studies" — a highly influential school of colonial and post-colonial history. I particularly enjoyed getting back to some of the foundational issues that I had to neglect somewhat at Cambridge.

While my thesis was quietly "maturing" with the mere passage of time (to be worthy of the 3-year rule), I took the liberty of submitting it for a competitive Prize Fellowship at Trinity College. Since, luckily, I also got elected, I then had to choose between continuing in Calcutta and going back to Cambridge. I split the time, and returned to Cambridge somewhat earlier than I had planned. The Prize Fellowship gave me four years of freedom to do anything I liked (no questions asked), and I took the radical decision of studying philosophy in that period. I had always been interested in logic and in epistemology, but soon got involved in moral and political philosophy as well (they related closely to my older concerns about democracy and equity).

The broadening of my studies into philosophy was important for me not just because some of my main areas of interest in economics relate quite closely to philosophical disciplines (for example, social choice theory makes intense use of mathematical logic and also draws on moral philosophy, and so does the study of inequality and deprivation), but also because I found philosophical studies very rewarding on their own. Indeed, I went on to write a number of papers in philosophy, particularly in epistemology, ethics and political philosophy. While I am interested both in economics and in philosophy, the union of my interests in the two fields far exceeds their intersection. When, many years later, I had the privilege of working with some major philosophers (such as John Rawls, Isaiah Berlin, Bernard Williams, Ronald Dworkin, Derek Parfit, Thomas Scanlon, Robert Nozick, and others), I felt very grateful to Trinity for having given me the opportunity as well as the courage to get into exacting philosophy.

Delhi School of Economics

During 1960–61, I visited M.I.T., on leave from Trinity College, and found it a great relief to get away from the rather sterile debates that the contending armies were fighting in Cambridge. I benefited greatly from many conversations with Paul Samuelson, Robert Solow, Franco Modigliani, Norbert Wiener, and others that made M.I.T. such an inspiring place. A summer visit to Stanford added to my sense of breadth of economics as a subject. In 1963, I decided to leave Cambridge altogether, and went to Delhi, as Professor of Economics at the Delhi School of Economics and at the University of Delhi. I taught in Delhi until 1971. In many ways this was the most intellectually challenging period of my academic life. Under the leadership of K. N. Raj, a remarkable applied economist who was already in Delhi, we made an attempt to build an advanced school of economics there. The Delhi School was already a good centre for economic study (drawing on the work of V. K. R. V. Rao, B. N. Ganguli, P. N. Dhar, Khaleq Naqvi, Dharm Narain, and many others, in addition to Raj), and a number of new economists joined, including Sukhamoy Chakravarty, Jagdish Bhagwati, A. L. Nagar, Manmohan Singh, Mrinal Datta Chaudhuri, Dharma Kumar, Raj Krishna, Ajit Biswas, K. L. Krishna, Suresh Tendulkar, and others. (Delhi School of Economics also had some leading social anthropologists, such as M. N. Srinivas, Andre Beteille, Baviskar, Veena Das, and major historians such as Tapan Ray Chaudhuri, whose work enriched the social sciences in general.) By the time I left Delhi in 1971 to join the London School of Economics, we had jointly succeeded in making the Delhi School the pre-eminent centre of education in economics and the social sciences, in India.

Regarding research, I plunged myself full steam into social choice theory in the dynamic intellectual atmosphere of Delhi University. My interest in the subject was consolidated during a one-year visit to Berkeley in 1964–65, where I not only had the chance to study and teach some social choice theory, but also had the unique opportunity of observing some practical social choice in the form of student activism in the "free speech movement." An initial difficulty in pursuing social choice at the Delhi School was that while I had the freedom to do what I liked, I did not, at first, have anyone who was interested in the subject as a formal discipline. The solution, of course, was to have students take an interest in the subject.

This happened with a bang with the arrival of a brilliant student, Prasanta Pattanaik, who did a splendid thesis on voting theory, and later on, also did joint work with me (adding substantially to the reach of what I was trying to do). Gradually, a sizeable and technically excellent group of economists interested in social choice theory emerged at the Delhi School.

Social choice theory related importantly to a more widespread interest in aggregation in economic assessment and policy making (related to poverty, inequality, unemployment, real national income, living standards). There was a great reason for satisfaction in the fact that a number of leading social choice theorists (in addition to Prasanta Pattanaik) emanated from the Delhi School, including Kaushik Basu and Rajat Deb (who also studied with me at the London School of Economics after I moved there), and Bhaskar Dutta and Manimay Sengupta, among others. There were other students who were primarily working in other areas (this applies to Basu as well), but whose work and interests were influenced by the strong current of social choice theory at the Delhi School (Nanak Kakwani is a good example of this).

In my book, *Collective Choice and Social Welfare*, published in 1970, I made an effort to take on overall view of social choice theory. There were a number of analytical findings to report, but despite the presence of many "trees" (in the form of particular technical results), I could not help looking anxiously for the forest. I had to come back again to the old general question that had moved me so much in my teenage years at Presidency College: Is reasonable social choice at all possible given the differences between one person's preferences (including interests and judgments) and another's (indeed, as Horace noted a long time ago, there may be "as many preferences as there are people")?

The work underlying *Collective Choice and Social Welfare* was mostly completed in Delhi, but I was much helped in giving it a final shape by a joint course on "social justice" I taught at Harvard with Kenneth Arrow and John Rawls, both of whom were wonderfully helpful in giving me their assessments and suggestions. The joint course was, in fact, quite a success both in getting many important issues discussed, and also in involving a remarkable circle of participants (who were sitting in as "auditors"), drawn from the established economists and philosophers in the Harvard region. (It was also quite well-known outside the campus: I was asked by a neighbour in a plane journey to San Francisco whether, as a teacher at Harvard, I had heard of an "apparently interesting" course taught by "Kenneth Arrow, John Rawls, and some unknown guy.")

There was another course I taught jointly, with Stephen Marglin and Prasanta Pattanaik (who too had come to Harvard), which was concerned with development as well as policy making. This nicely supplemented my involvements in pure social choice theory (in fact, Marglin and Pattanaik were both very interested in examining the connection between social choice theory and other areas in economics).

From Delhi to London and Oxford

I left Delhi, in 1971, shortly after *Collective Choice and Social Welfare* was published in 1970. My wife, Nabaneeta Dev, with whom I have two children (Antara and Nandana), had constant trouble with her health in Delhi (mainly from asthma). London might have suited her better, but, as it happens, the marriage broke up shortly after we went to London.

Nabaneeta is a remarkably successful poet, literary critic and writer of novels and short stories (one of the most celebrated authors in contemporary Bengali literature), which she has combined, since our divorce, with being a University Professor at Jadavpur University in Calcutta. I learned many things from her, including the appreciation of poetry from an "internal" perspective. She had worked earlier on the distinctive style and composition of epic poetry, including the Sanskrit epics (particularly the *Ramayana*), and this I had got very involved in. Nabaneeta's parents were very well-known poets as well, and she seems to have borne her celebrity status — and the great many recognitions that have come her way — with unaffected approachability and warmth. She had visits from an unending stream of literary fans, and I understand, still does. (On one occasion, arrived a poet with a

hundred new poems, with the declared intention of reading them aloud
to her, to get her critical judgement, but since she was out, he said that
he would instead settle for reading them to me. When I pleaded that I
lacked literary sophistication, I was assured by the determined poet:
"That is just right; I would like to know how the common man may
react to my poetry." The common man, I am proud to say, reacted with
appropriate dignity and self-control.)

When we moved to London, I was also going through some serious
medical problems. In early 1952, at the age of 18 (when I was an
undergraduate at Presidency College), I had cancer of the mouth, and it
had been dealt with by a severe dose of radiation in a rather primitive
Calcutta hospital. This was only seven years after Hiroshima and
Nagasaki, and the long-run effects of radiation were not much
understood. The dose of radiation I got may have cured the cancer, but
it also killed the bones in my hard palate. By 1971, it appeared that I
had either a recurrence of the cancer, or a severe case of bone necrosis.
The first thing I had to do on returning to England was to have a
serious operation, without knowing whether it would be merely plastic
surgery to compensate for the necrosis (a long and complicated
operation in the mouth, but no real threat to survival), or much more
demandingly, a fresh round of efforts at cancer eradication.

After the long operation (it had lasted nearly seven hours) when I
woke up from the heavy anaesthesia, it was four o'clock in the morning.
As a person with much impatience, I wanted to know what the surgeon
had found. The nurse on duty said she was not allowed to tell me
anything: "You must wait for the doctors to come at nine." This created
some tension (I wanted to know what had emerged), which the nurse
noticed. I could see that she was itching to tell me something: indeed
(as I would know later) to tell me that no recurrence of cancer had been
detected in the frozen-section biopsy that had been performed, and
that the long operation was mainly one of reconstruction of the palate
to compensate for the necrosis. She ultimately gave in, and chose an
interesting form of communication, which I found quite striking
(as well as kind). "You know," she said, "they were praising you very
much!" It then dawned on me that not having cancer can be a subject
for praise. Indeed lulled by praise, I went quietly back to my post-

operative sleep. In later years, when I would try to work on judging the goodness of a society by the quality of health of the people, her endorsement of my praiseworthiness for being cancer-free would serve as a good reference point!

The intellectual atmosphere at the LSE in particular and in London in general was most gratifying, with a dazzling array of historians, economists, sociologists and others. It was wonderful to have the opportunity of seeing Eric Hobsbawm (the great historian) and his wife Marlene very frequently and to interact regularly with Frank and Dorothy Hahn, Terence and Dorinda Gorman, and many others. Our small neighbourhood in London (Bartholomew estate, within the Kentish Town) itself offered wonderful company of intellectual and artistic creativity and political involvement. Even after I took an Oxford job (Professor of Economics, 1977–80, Drummond Professor of Political Economy, 1980–87) later on, I could not be budged from living in London.

As I settled down at the London School of Economics in 1971, I resumed my work on social choice theory. Again, I had excellent students at LSE, and later on at Oxford. In addition to Kaushik Basu and Rajat Deb (who had come from Dehli), other students such as Siddiq Osmani, Ben Fine, Ravi Kanbur, Carl Hamilton, John Wriglesworth, David Kelsey, Yasumi Matsumoto, Jonathan Riley, produced distinguished Ph.D. theses on a variety of economic and social choice problems. It made me very proud that many of the results that became standard in social choice theory and welfare economics had first emerged in these Ph.D. theses.

I was also fortunate to have colleagues who were working on serious social choice problems, including Peter Hammond, Charles Blackorby, Kotaro Suzumura, Geoffrey Heal, Gracieda Chichilnisky, Ken Binmore, Wulf Gaertner, Eric Maskin, John Muellbauer, Kevin Roberts, Susan Hurley, at LSE or Oxford, or neighbouring British universities. (I also learned greatly from conversations with economists who were in other fields, but whose works were of great interest to me, including Sudhir Anand, Tony Atkinson, Christopher Bliss, Meghnad Desai, Terence Gorman, Frank Hahn, David Hendry, Richard Layard, James Mirrlees, John Muellbauer, Steve Nickel, among others.) I also had the opportunity of collaboration with social choice theorists elsewhere, such as Claude d'Aspremont and Louis Gevers in Belgium, Koichi Hamada and Ken-ichi Inada in Japan (joined later by Suzumura when he returned there), and many others in America, Canada, Israel, Australia, Russia, and elsewhere). There were many new formal results and informal understandings that emerged in these works, and the gloom of "impossibility results" ceased to be the only prominent theme in the field. The 1970s were probably the golden years of social choice theory across the world. Personally, I had the sense of having a ball.

From Social Choice to Inequality and Poverty

The constructive possibilities that the new literature on social choice produced directed us immediately to making use of available statistics for a variety of economic and social appraisals: measuring economic inequality, judging poverty, evaluating projects, analyzing unemployment, investigating the principles and implications of liberty and rights, assessing gender inequality, and so on. My work on inequality was much inspired and stimulated by that of Tony Atkinson. I also worked for a while with Partha Dasgupta and David Starrett on measuring inequality (after having worked with Dasgupta and Stephen Marglin on project evaluation), and later, more extensively, with Sudhir Anand and James Foster.

My own interests gradually shifted from the pure theory of social choice to more "practical" problems. But I could not have taken them on without having some confidence that the practical exercises to be undertaken were also foundationally secure (rather than implicitly harbouring incongruities and impossibilities that could be exposed on deeper analytical probing). The progress of the pure theory of social choice with an expanded informational base was, in this sense, quite crucial for my applied work as well.

In the reorientation of my research, I benefited greatly from discussions with my wife, Eva Colorni, with whom I lived from 1973 onwards. Her critical standards were extremely exacting, but she also wanted to encourage me to work on issues of practical moment. Her personal background involved a fine mixture of theory and practice, with an Italian Jewish father (Eugenio Colorni was an academic philosopher and a hero of the Italian resistance who was killed by the fascists in Rome shortly before the Americans got there), a Berlinite Jewish mother (Ursula Hirschman was herself a writer and the brother of the great development economist, Albert Hirschman), and a stepfather who as a statesman had been a prime mover in uniting Europe (Altiero Spinelli was the founder of the "European Federalist movement," wrote its "Manifesto" from prison in 1941, and officially established the new movement, in the company of Eugenio Colorni, in Milan in 1943). Eva herself had studied law, philosophy and economics (in Pavia and in Delhi), and lectured at the City of London Polytechnic (now London Guildhall University). She was deeply humane (with a great passion for social justice) as well as fiercely rational (taking no theory for granted, subjecting each to reasoned assessment and scrutiny). She exercised a great influence on the standards and reach that I attempted to achieve in my work (often without adequate success).

Eva was very supportive of my attempt to use a broadened framework of social choice theory in a variety of applied problems: to assess poverty; to evaluate inequality; to clarify the nature of relative deprivation; to develop distribution-adjusted national income measures; to clarify the penalty of unemployment; to analyze violations of personal liberties and basic rights; and to characterize gender disparities and women's relative disadvantage. The results were mostly published in journals in the 1970s and early 1980s, but gathered together in two collections of articles (*Choice, Welfare and Measurement* and *Resources, Values and Development*, published, respectively, in 1982 and 1984).

The work on gender inequality was initially confined to analyzing available statistics on the male–female differential in India (I had a joint paper with Jocelyn Kynch on "Indian Women: Well-being and Survival" in 1982), but gradually moved to international comparisons (*Commodities and Capabilities*, 1985) and also to some general theory (*Gender and Cooperative Conflicts*, 1990). The theory drew both on empirical analysis of published statistics across the world, but also of data I freshly collected in India in the spring of 1983, in collaboration with Sunil Sengupta, comparing boys and girls from birth to age 5. (We weighed and studied every child in two largish villages in West Bengal; I developed some expertise in weighing protesting children, and felt quite proud of my accomplishment when, one day, my research assistant phoned me with a request to take over from her 'the job of weighing a child "who bites every hand within the reach of her teeth." I developed some vanity in being able to meet the challenge at the "biting end" of social choice research.)

Poverty, Famines and Deprivation

From the mid-1970s, I also started work on the causation and prevention of famines. This was initially done for the World Employment Programme of the International Labour Organization, for which my 1981 book *Poverty and Famines* was written. (Louis Emmerij who led the programme took much personal interest in the work I was trying to do on famines.) I attempted to see famines as broad "economic" problems (concentrating on how people can buy food, or otherwise get entitled to it), rather than in terms of the grossly undifferentiated picture of aggregate food supply for the economy as a whole. The work was carried on later (from the middle of 1980s) under the auspices of the World Institute of Development Economics Research (WIDER) in Helsinki, which was imaginatively directed by Lal Jayawardena (an old friend who, as I noted earlier, had also been a contemporary of mine at Cambridge in the 1950s). Siddiq Osmani, my ex-student, ably led the programme on hunger and deprivation at WIDER. I also worked closely with Martha Nussbaum on the cultural side of the programme, during 1987–89.

By the mid-1980s, I was collaborating extensively with Jean Drèze, a young Belgian economist of extraordinary skill and remarkable dedication. My understanding of hunger and deprivation owes a great deal to his insights and investigations, and so does my recent work on development, which has been mostly done jointly with him. Indeed, my collaboration with Jean has been extremely fruitful for me, not only because I have learned so much from his imaginative initiatives and insistent thoroughness, but also because it is hard to beat an arrangement for joint work whereby Jean does most of the work whereas I get a lot of the credit.

While these were intensely practical matters, I also got more and more involved in trying to understand the nature of individual advantage in terms of the substantive freedoms that different persons respectively enjoy, in the form of the capability to achieve valuable things. If my work in social choice theory was initially motivated by a desire to overcome Arrow's pessimistic picture by going beyond his limited informational base, my work on social justice based on individual freedoms and capabilities was similarly motivated by an aspiration to learn from, but go beyond, John Rawls's elegant theory of justice, through a broader use of available information. My intellectual life has been much influenced by the contributions as well as the wonderful helpfulness of both Arrow and Rawls.

Harvard and Beyond

In the late 1980s, I had reason to move again from where I was. My wife, Eva, developed a difficult kind of cancer (of the stomach), and died quite suddenly in 1985. We had young children (Indrani and Kabir — then 10 and 8 respectively), and I wanted to take them away to another country, where they would not miss their mother constantly. The liveliness of America appealed to us as an alternative location, and I took the children with me to "taste" the prospects in the American universities that made me an offer.

Indrani and Kabir rapidly became familiar with several campuses (Stanford, Berkeley, Yale, Princeton, Harvard, UCLA, University of Texas at Austin, among them), even though their knowledge of America outside academia remained rather limited. (They particularly enjoyed visiting their grand uncle and aunt, Albert and Sarah Hirschman, at the Institute for Advanced Study in Princeton; as a Trustee of the Institute, visits to Princeton were also very pleasurable occasions for me.) I guess I was, to some extent, imposing my preference for the academic climate on the children, by confining the choice to universities only, but I did not really know what else to do. However, I must confess that I worried a little when I overheard my son Kabir, then nine years old, responding to a friendly American's question during a plane journey as to whether he knew Washington, D.C. "Is that city," I heard Kabir say, "closer to Palo Alto or to New Haven?"

We jointly chose Harvard, and it worked out extremely well. My colleagues in economics and philosophy were just superb, some of whom I knew well from earlier on (including John Rawls and Tim Scanlon in philosophy, and Zvi Griliches, Dale Jorgenson, Janos Kornai, Stephen Marglin in economics), but there were also others whom I came to know after arriving at Harvard. I greatly enjoyed teaching regular joint courses with Robert Nozick and Eric Maskin, and also on occasions, with John Rawls and Thomas Scanlon (in philosophy) and with Jerry Green, Stephen Marglin and David Bloom (in economics). I could learn also from academics in many other fields as well, not least at the Society of Fellows where I served as a Senior Fellow for nearly a decade. Also, I was again blessed with wonderful students in economics, philosophy, public health and government, who did excellent theses, including Andreas Papandreou (who moved with me from Oxford to Harvard, and did a major book on externality and the environment), Tony Laden (who, among many other things, clarified the game-theoretic structure of Rawlsian theory of justice), Stephan Klasen (whose work on gender inequality in survival is possibly the most definitive work in this area), Felicia Knaul (who worked on street children and the economic and social challenges they face), Jennifer Ruger (who substantially advance the understanding of health as a public policy concern), and indeed many others with whom I greatly enjoyed working.

The social choice problems that had bothered me earlier on were by now more analyzed and understood, and I did have, I thought, some understanding of the demands of fairness, liberty and equality. To get firmer understanding of all this, it was necessary to pursue further the search for an adequate characterization of individual advantage. This had been the subject of my Tanner Lectures on Human Values at Stanford in 1979 (published as a paper, "Equality of What?" in 1980) and in a more empirical form, in a second set of Tanner Lectures at Cambridge in 1985 (published in 1987 as a volume of essays, edited by Geoffrey Hawthorne, with contributions by Bernard Williams, Ravi

Kanbur, John Muellbauer, and Keith Hart). The approach explored sees individual advantage not merely as opulence or utility, but primarily in terms of the lives people manage to live and the freedom they have to choose the kind of life they have reason to value. The basic idea here is to pay attention to the actual "capabilities" that people end up having. The capabilities depend both on our physical and mental characteristics as well as on social opportunities and influences (and can thus serve as the basis not only of assessment of personal advantage but also of efficiency and equity of social policies). I was trying to explore this approach since my Tanner Lectures in 1979; there was a reasonably ambitious attempt at linking theory to empirical exercises in my book *Commodities and Capabilities*, published in 1985. In my first few years at Harvard, I was much concerned with developing this perspective further.

The idea of capabilities has strong Aristotelian connections, which I came to understand more fully with the help of Martha Nussbaum, a scholar with a remarkably extensive command over classical philosophy as well as contemporary ethics and literary studies. I learned a great deal from her, and we also collaborated in a number of studies during 1987–89, including in a collection of essays that pursued this approach in terms of philosophical as well as economic reasoning (*Quality of Life* was published in 1993, but the essays were from a conference at WIDER in Helsinki in 1988).

During my Harvard years up to about 1991, I was much involved in analyzing the overall implications of this perspective on welfare economics and political philosophy (this is reported in my book, *Inequality Reexamined*, published in 1992). But it was also very nice to get involved in some new problems, including the characterization of rationality, the demands of objectivity, and the relation between facts and values. I used the old technique of offering courses on them (sometimes jointly with Robert Nozick) and through that learning as much as I taught. I started taking an interest also in health equity (and in public health in particular, in close collaboration with Sudhir Anand), a challenging field of application for concepts of equity and justice. Harvard's ample strength in an immense variety of subjects gives one scope for much freedom in the choice of work and of colleagues to talk to, and the high quality of the students was a total delight as well. My work on inequality in terms of variables other than incomes was also helped by the collaboration of Angus Deaton and James Foster.

It was during my early years at Harvard that my old friend, Mahbub ul Haq, who had been a fellow student at Cambridge (and along with his wife, Bani, a very old and close friend), returned back into my life in a big way. Mahbub's professional life had taken him from Cambridge to Yale, then back to his native Pakistan, with intermediate years at the World Bank. In 1989 he was put in charge, by the United Nations Development Programme (UNDP), of the newly planned "Human Development Reports." Mahbub insisted that I work with him to help develop a broader informational approach to the assessment of development. This I did with great delight, partly because of the exciting nature of the work, but also because of the opportunity of working closely with such an old and wonderful friend. *Human Development Reports* seem to have received a good deal of attention in international circles, and Mahbub was very successful in broadening the informational basis of the assessment of development. His sudden death in 1998 has robbed the world of one of the leading practical reasoners in the world of contemporary economics.

India and Bangladesh

What about India? While I have worked abroad since 1971, I have constantly retained close connections with Indian universities, I have, of course, a special relation with Delhi University, where I have been an honorary professor since leaving my full-time job there in 1971, and I use this excuse to subject Delhi students to lectures whenever I get a chance. For various reasons — personal as well as academic — the peripatetic life seems to suit me, in this respect. After my student days in Cambridge in 1953–56, I guess I have never been away from India for more than six months at a time. This - combined with my remaining exclusively an Indian citizen — gives me, I think, some entitlement to speak on Indian public affairs, and this remains a constant involvement.

It is also very engaging — and a delight — to go back to Bangladesh as often as I can, which is not only my old home, but also where some of my closest friends and collaborators live and work. This includes Rehman Sobhan to whom I have been very close from my student days (he remains as sceptical of formal economics and its reach as he was in the early 1950s), and also Anisur Rehman (who is even more sceptical), Kamal Hossain, Jamal Islam, Mushairaf Hussain, among many others, who are all in Bangladesh.

When the Nobel award came my way, it also gave me an opportunity to do something immediate and practical about my old obsessions, including literacy, basic health care and gender equity, aimed specifically at India and Bangladesh. The Pratichi Trust, which I have set up with the help of some of the prize money, is, of course, a small effort compared with the magnitude of these problems. But it is nice to re-experience something of the old excitement of running evening schools, more than fifty years ago, in villages near Santiniketan.

From Campus to Campus

As far as my principal location is concerned, now that my children have grown up, I could seize the opportunity to move back to my old Cambridge college, Trinity. I accepted the offer of becoming Master of the College from January 1998 (though I have not cut my connections with Harvard altogether). The reasoning was not independent of the fact that Trinity is not only my old college where my academic life really began, but it also happens to be next door to King's, where my wife, Emma Rothschild, is a Fellow, and Director of the Centre for History and Economics. Her book on Adam Smith also takes on the hard task of reinterpreting the European Enlightenment. It so happens that one principal character in this study is Condorcet, who was also one of the originators of social choice theory, which is very pleasing (and rather useful as well).

Emma too is a convinced academic (a historian and an economist), and both her parents had long connections with Cambridge and with the University. Between my four children, and the two of us, the universities that the Sen family has encountered include Calcutta University, Cambridge University, Jadavpur University, Delhi University, L.S.E., Oxford University, Harvard University, M.I.T., University of

California at Berkeley, Stanford University, Cornell University, Smith College, Wesleyan University, among others. Perhaps one day we can jointly write an illustrated guide to the universities.

I end this essay where I began — at a university campus. It is not quite the same at 65 as it was at 5. But it is not so bad even at an older age (especially, as Maurice Chevalier has observed, "considering the alternative"). Nor are university campuses quite as far removed from life as is often presumed. Robert Goheen has remarked, "if you feel that you have both feet planted on level ground, then the university has failed you." Right on. But then who wants to be planted on ground? There are places to go.

✲

Amartya Sen receiving his Prize from the hands of His Majesty the King.

Rabindranath TAGORE
Photo courtesy of the Nobel Foundation.

Tagore and His India[*]

❇

Amartya Sen

Voice of Bengal

Rabindranath Tagore, who died in 1941 at the age of eighty, is a towering figure in the millennium-old literature of Bengal. Anyone who becomes familiar with this large and flourishing tradition will be impressed by the power of Tagore's presence in Bangladesh and in India. His poetry as well as his novels, short stories, and essays are very widely read, and the songs he composed reverberate around the eastern part of India and throughout Bangladesh.

In contrast, in the rest of the world, especially in Europe and America, the excitement that Tagore's writings created in the early years of the twentieth century has largely vanished. The enthusiasm with which his work was once greeted was quite remarkable. *Gitanjali*, a selection of his poetry for which he was awarded the Nobel Prize in Literature in 1913, was published in English translation in London in March of that year, and had been reprinted ten times by November, when the award was announced. But he is not much read now in the

---

[*] First published August 28, 2001. Courtesy of *The New York Review*.

West, and already by 1937, Graham Greene was able to say: "As for Rabindranath Tagore, I cannot believe that anyone but Mr. Yeats can still take his poems very seriously."

The Mystic

The contrast between Tagore's commanding presence in Bengali literature and culture, and his near-total eclipse in the rest of the world, is perhaps less interesting than the distinction between the view of Tagore as a deeply relevant and many-sided contemporary thinker in Bangladesh and India, and his image in the West as a repetitive and remote spiritualist. Graham Greene had, in fact, gone on to explain that he associated Tagore "with what Chesterton calls 'the bright pebbly eyes' of the Theosophists." Certainly, an air of mysticism played some part in the "selling" of Rabindranath Tagore to the West by Yeats, Ezra Pound, and his other early champions. Even Anna Akhmatova, one of Tagore's few later admirers (who translated his poems into Russian in the mid-1960s), talks of "that mighty flow of poetry which takes its strength from Hinduism as from the Ganges, and is called Rabindranath Tagore."

Confluence of Cultures

Rabindranath did come from a Hindu family — one of the landed gentry who owned estates mostly in what is now Bangladesh. But whatever wisdom there might be in Akhmatova's invoking of Hinduism and the Ganges, it did not prevent the largely Muslim citizens of Bangladesh from having a deep sense of identity with Tagore and his ideas. Nor did it stop the newly independent Bangladesh from choosing one of Tagore's songs — the "Amar Sonar Bangla" which means "my golden Bengal" — as its national anthem. This must be very confusing to those who see the contemporary world as a "clash of civilizations" — with "the Muslim civilization," "the Hindu civilization," and "the Western civilization," each forcefully confronting the others. They would also be confused by Rabindranath Tagore's own description of his Bengali family as the product of "a confluence of three cultures: Hindu, Mohammedan, and British".[1]

Rabindranath's grandfather, Dwarkanath, was well known for his command of Arabic and Persian, and Rabindranath grew up in a family atmosphere in which a deep knowledge of Sanskrit and ancient Hindu texts was combined with an understanding of Islamic traditions as well as Persian literature. It is not so much that Rabindranath tried to produce — or had an interest in producing — a "synthesis" of the different religions (as the great Moghul emperor Akbar tried hard to achieve) as that his outlook was persistently non-sectarian, and his writings — some two hundred books — show the influence of different parts of the Indian cultural background as well as of the rest of the world.[2]

Abode of Peace

Most of his work was written at Santiniketan (Abode of Peace), the small town that grew around the school he founded in Bengal in 1901, and he not only conceived there an imaginative and innovative system of education, but through his writings and his influence on students and teachers, he was able to use the school as a base from which he could take a major part in India's social, political, and cultural movements.

The profoundly original writer, whose elegant prose and magical poetry Bengali readers know well, is not the sermonizing spiritual guru admired — and then rejected — in London. Tagore was not only an immensely versatile poet; he was also a great short story writer, novelist, playwright, essayist, and composer of songs, as well as a talented painter whose pictures, with their mixture of representation and abstraction, are only now beginning to receive the acclaim that they have long deserved. His essays, moreover, ranged over literature, politics, culture, social change, religious beliefs, philosophical analysis, international relations, and much else. The coincidence of the fiftieth anniversary of Indian independence with the publication of a selection of Tagore's letters by Cambridge University Press,[3] brought Tagore's ideas and reflections to the fore, which makes it important to examine what kind of leadership in thought and understanding he provided in the Indian subcontinent in the first half of this century.

Gandhi and Tagore

Since Rabindranath Tagore and Mohandas Gandhi were two leading Indian thinkers in the twentieth century, many commentators have tried to compare their ideas. On learning of Rabindranath's death, Jawaharlal Nehru, then incarcerated in a British jail in India, wrote in his prison diary for August 7, 1941:

> *"Gandhi and Tagore. Two types entirely different from each other, and yet both of them typical of India, both in the long line of India's great men ... It is not so much because of any single virtue but because of the tout ensemble, that I felt that among the world's great men today Gandhi and Tagore were supreme as human beings. What good fortune for me to have come into close contact with them."*

Romain Rolland was fascinated by the contrast between them, and when he completed his book on Gandhi, he wrote to an Indian academic, in March 1923: "I have finished my *Gandhi*, in which I pay tribute to your two great river-like souls, overflowing with divine spirit, Tagore and Gandhi." The following month, he recorded in his diary an account of some of the differences between Gandhi and Tagore written by Reverend C. F. Andrews, the English clergyman and public activist who was a close friend of both men (and whose important role in Gandhi's life in South Africa as well as India is well portrayed in Richard Attenborough's film *Gandhi* [1982]). Andrews described to Rolland a discussion between Tagore and Gandhi, at which he was present, on subjects that divided them:

> *"The first subject of discussion was idols; Gandhi defended them, believing the masses incapable of raising themselves immediately to abstract ideas. Tagore cannot bear to see the*

people eternally treated as a child. Gandhi quoted the great
things achieved in Europe by the flag as an idol; Tagore
found it easy to object, but Gandhi held his ground,
contrasting European flags bearing eagles, etc., with his
own, on which he has put a spinning wheel. The second
point of discussion was nationalism, which Gandhi
defended. He said that one must go through nationalism
to reach internationalism, in the same way that one must
go through war to reach peace."[4]

Tagore greatly admired Gandhi but he had many disagreements with him on a variety of subjects, including nationalism, patriotism, the importance of cultural exchange, the role of rationality and of science, and the nature of economic and social development. These differences, I shall argue, have a clear and consistent pattern, with Tagore pressing for more room for reasoning, and for a less traditionalist view, a greater interest in the rest of the world, and more respect for science and for objectivity generally.

Rabindranath knew that he could not have given India the political leadership that Gandhi provided, and he was never stingy in his praise for what Gandhi did for the nation (it was, in fact, Tagore who popularized the term "Mahatma" — great soul — as a description of Gandhi). And yet each remained deeply critical of many things that the other stood for. That Mahatma Gandhi has received incomparably more attention outside India and also within much of India itself makes it important to understand "Tagore's side" of the Gandhi–Tagore debates.

In his prison diary, Nehru wrote: "Perhaps it is as well that [Tagore] died now and did not see the many horrors that are likely to descend in increasing measure on the world and on India. He had seen enough and he was infinitely sad and unhappy." Toward the end of his life, Tagore was indeed becoming discouraged about the state of India, especially as its normal burden of problems, such as hunger and poverty, was being supplemented by politically organized incitement to "communal" violence between Hindus and Muslims. This conflict would lead in

1947, six years after Tagore's death, to the widespread killing that took place during partition; but there was much gore already during his declining days. In December 1939, he wrote to his friend Leonard Elmhirst, the English philanthropist and social reformer who had worked closely with him on rural reconstruction in India (and who had gone on to found the Dartington Hall Trust in England and a progressive school at Dartington that explicitly invoked Rabindranath's educational ideals):[5]

> *"It does not need a defeatist to feel deeply anxious about the future of millions who, with all their innate culture and their peaceful traditions are being simultaneously subjected to hunger, disease, exploitations foreign and indigenous, and the seething discontents of communalism."*

How would Tagore have viewed the India of today? Would he see progress there, or wasted opportunity, perhaps even a betrayal of its promise and conviction? And, on a wider subject, how would he react to the spread of cultural separatism in the contemporary world?

East and West

Given the vast range of his creative achievements, perhaps the most astonishing aspect of the image of Tagore in the West is its narrowness; he is recurrently viewed as "the great mystic from the East," an image with a putative message for the West, which some would welcome, others dislike, and still others find deeply boring. To a great extent this Tagore was the West's own creation, part of its tradition of message-seeking from the East, particularly from India, which — as Hegel put it — had "existed for millennia in the imagination of the Europeans."[6] Friedrich Schlegel, Schelling, Herder, and Schopenhauer were only a few of the thinkers who followed the same pattern. They theorized, at first, that India was the source of superior wisdom. Schopenhauer at one stage even argued that the New Testament "must somehow be of Indian origin: this is attested by its completely Indian ethics, which transforms morals into asceticism, its pessimism, and its avatar," in "the person of Christ." But then they rejected their own theories with great vehemence, sometimes blaming India for not living up to their unfounded expectations.

We can imagine that Rabindranath's physical appearance — handsome, bearded, dressed in non-Western clothes — may, to some extent, have encouraged his being seen as a carrier of exotic wisdom. Yasunari Kawabata, the first Japanese Nobel Laureate in Literature, treasured memories from his middle-school days of "this sage-like poet":

> *His white hair flowed softly down both sides of his*
> *forehead; the tufts of hair under the temples also were*
> *long like two beards, and linking up with the hair on*
> *his cheeks, continued into his beard, so that he gave an*
> *impression, to the boy I was then, of some ancient*
> *Oriental wizard.*[7]

That appearance would have been well-suited to the selling of Tagore in the West as a quintessentially mystical poet, and it could have made it somewhat easier to pigeonhole him. Commenting on Rabindranath's appearance, Frances Cornford told William Rothenstein, "I can now imagine a powerful and gentle Christ, which I never could before." Beatrice Webb, who did not like Tagore and resented what she took to be his "quite obvious dislike of all that the Webbs stand for" (there is, in fact, little evidence that Tagore had given much thought to this subject), said that he was "beautiful to look at" and that "his speech has the perfect intonation and slow chant-like moderation of the dramatic saint." Ezra Pound and W. B. Yeats, among others, first led the chorus of adoration in the Western appreciation of Tagore, and then soon moved to neglect and even shrill criticism. The contrast between Yeats's praise of his work in 1912 ("These lyrics ... display in their thought a world I have dreamed of all my life long," "the work of a supreme culture") and his denunciation in 1935 ("Damn Tagore") arose partly from the inability of Tagore's many-sided writings to fit into the narrow box in which Yeats wanted to place — and keep — him. Certainly,

Tagore did write a huge amount, and published ceaselessly, even in English (sometimes in indifferent English translation), but Yeats was also bothered, it is clear, by the difficulty of fitting Tagore's later writings into the image Yeats had presented to the West. Tagore, he had said, was the product of "a whole people, a whole civilization, immeasurably strange to us," and yet "we have met our own image, ... or heard, perhaps for the first time in literature, our voice as in a dream."[8]

Yeats did not totally reject his early admiration (as Ezra Pound and several others did), and he included some of Tagore's early poems in *The Oxford Book of Modern Verse*, which he edited in 1936. Yeats also had some favorable things to say about Tagore's prose writings. His censure of Tagore's later poems was reinforced by his dislike of Tagore's own English translations of his work ("Tagore does not know English, no Indian knows English," Yeats explained), unlike the English version of *Gitanjali* which Yeats had himself helped to prepare. Poetry is, of course, notoriously difficult to translate, and anyone who knows Tagore's poems in their original Bengali cannot feel satisfied with any of the translations (made with or without Yeats's help). Even the translations of his prose works suffer, to some extent, from distortion. E. M. Forster noted, in a review of a translation of one of Tagore's great Bengali novels, *The Home and the World*, in 1919: "The theme is so beautiful," but the charms have "vanished in translation," or perhaps "in an experiment that has not quite come off."[9]

Tagore himself played a somewhat bemused part in the boom and bust of his English reputation. He accepted the extravagant praise with much surprise as well as pleasure, and then received denunciations with even greater surprise, and barely concealed pain. Tagore was sensitive to criticism, and was hurt by even the most far-fetched accusations, such as the charge that he was getting credit for the work of Yeats, who had "rewritten" *Gitanjali*. (This charge was made by a correspondent for *The Times*, Sir Valentine Chirol, whom E. M. Forster once described as "an old Anglo-Indian reactionary hack.") From time to time Tagore also protested the crudity of some of his overexcited advocates. He wrote to C. F. Andrews in 1920: "These people ... are like drunkards who are afraid of their lucid intervals."

God and Others

Yeats was not wrong to see a large religious element in Tagore's writings. He certainly had interesting and arresting things to say about life and death. Susan Owen, the mother of Wilfred Owen, wrote to Rabindranath in 1920, describing her last conversations with her son before he left for the war which would take his life. Wilfred said goodbye with "those wonderful words of yours — beginning at 'When I go from hence, let this be my parting word.'" When Wilfred's pocket notebook was returned to his mother, she found "these words written in his dear writing — with your name beneath."

The idea of a direct, joyful, and totally fearless relationship with God can be found in many of Tagore's religious writings, including the poems of *Gitanjali*. From India's diverse religious traditions he drew many ideas, both from ancient texts and from popular poetry. But "the bright pebbly eyes of the Theosophists" do not stare out of his verses. Despite the archaic language of the original translation of *Gitanjali*, which did not, I believe, help to preserve the simplicity of the original, its elementary humanity comes through more clearly than any complex and intense spirituality:

> *Leave this chanting and singing and telling of beads!*
> *Whom dost thou worship in this lonely dark corner of a*
> *temple with doors all shut?*
> *Open thine eyes and see thy God is not before thee!*
> *He is there where the tiller is tilling the hard ground and*
> *where the pathmaker is breaking stones.*
> *He is with them in sun and in shower, and his garment is*
> *covered with dust.*

An ambiguity about religious experience is central to many of Tagore's devotional poems, and makes them appeal to readers irrespective of their beliefs; but excessively detailed interpretation can ruinously strip away that ambiguity.[10] This applies particularly to his many poems which combine images of human love and those of pious devotion. Tagore writes:

*I have no sleep to-night. Ever and again I open my door
and look out on the darkness, my friend!
I can see nothing before me. I wonder where lies thy
path!
By what dim shore of the ink-black river, by what far
edge of the frowning forest, through what mazy depth of
gloom, art thou threading thy course to come to see me,
my friend?*

I suppose it could be helpful to be told, as Yeats hastens to explain, that "the servant or the bride awaiting the master's home-coming in the empty house" is "among the images of the heart turning to God." But in Yeats's considerate attempt to make sure that the reader does not miss the "main point," something of the enigmatic beauty of the Bengali poem is lost — even what had survived the antiquated language of the English translation. Tagore certainly had strongly held religious beliefs (of an unusually nondenominational kind), but he was interested in a great many other things as well and had many different things to say about them.

Some of the ideas he tried to present were directly political, and they figure rather prominently in his letters and lectures. He had practical, plainly expressed views about nationalism, war and peace, cross-cultural education, freedom of the mind, the importance of rational criticism, the need for openness, and so on. His admirers in the West, however, were tuned to the more otherworldly themes which had been emphasized by his first Western patrons. People came to his public lectures in Europe and America, expecting ruminations on grand, transcendental themes; when they heard instead his views on the way public leaders should behave, there was some resentment, particularly (as E. P. Thompson reports) when he delivered political criticism "at $700 a scold."

Reasoning in Freedom

For Tagore it was of the highest importance that people be able to live, and reason, in freedom. His attitudes toward politics and culture, nationalism and internationalism, tradition and modernity, can all be seen in the light of this belief.[11] Nothing, perhaps, expresses his values as clearly as a poem in *Gitanjali*:

> *Where the mind is without fear*
> *and the head is held high;*
> *Where knowledge is free;*
> *Where the world has not been*
> *broken up into fragments*
> *by narrow domestic walls; ...*
> *Where the clear stream of reason*
> *has not lost its way into the*
> *dreary desert sand of dead habit; ...*
> *Into that heaven of freedom,*
> *my Father, let my country awake.*

Rabindranath's qualified support for nationalist movements — and his opposition to the unfreedom of alien rule — came from this commitment. So did his reservations about patriotism, which, he argued, can limit both the freedom to engage ideas from outside "narrow domestic walls" and the freedom also to support the causes of people in other countries. Rabindranath's passion for freedom underlies his firm opposition to unreasoned traditionalism, which makes one a prisoner of the past (lost, as he put it, in "the dreary desert sand of dead habit").

Tagore illustrates the tyranny of the past in his amusing yet deeply serious parable "Kartar Bhoot" ("The Ghost of the Leader"). As the respected leader of an imaginary land is about to die, his panic-stricken followers request him to stay on after his death to instruct them on what to do. He consents. But his followers find their lives are full of rituals and constraints on everyday behavior and are not responsive to the world around them. Ultimately, they request the ghost of the leader to relieve them of his domination, when he informs them that he exists only in their minds.

Tagore's deep aversion to any commitment to the past that could not be modified by contemporary reason extended even to the alleged virtue of invariably keeping past promises. On one occasion when Mahatma Gandhi visited Tagore's school at Santiniketan, a young woman got him to sign her autograph book. Gandhi wrote: "Never make a promise in haste. Having once made it fulfill it at the cost of your life." When he saw this entry, Tagore became agitated. He wrote in the same book a short poem in Bengali to the effect that no one can be made "a prisoner forever with a chain of clay." He went on to conclude in English, possibly so that Gandhi could read it too, "Fling away your promise if it is found to be wrong."[12]

Tagore had the greatest admiration for Mahatma Gandhi as a person and as a political leader, but he was also highly skeptical of Gandhi's form of nationalism and his conservative instincts regarding the country's past traditions. He never criticized Gandhi personally. In the 1938 essay, "Gandhi the Man," he wrote:

> Great as he is as a politician, as an organizer, as a leader
> of men, as a moral reformer, he is greater than all these
> as a man, because none of these aspects and activities
> limits his humanity. They are rather inspired and sustained
> by it.

And yet there is a deep division between the two men. Tagore was explicit about his disagreement:

> We who often glorify our tendency to ignore reason,
> installing in its place blind faith, valuing it as spiritual,
> are ever paying for its cost with the obscuration of our mind
> and destiny. I blamed Mahatmaji for exploiting this
> irrational force of credulity in our people, which might have
> had a quick result [in creating] a superstructure, while
> sapping the foundation. Thus began my estimate of
> Mahatmaji, as the guide of our nation, and it is fortunate
> for me that it did not end there.

But while it "did not end there," that difference of vision was a powerful divider. Tagore, for example, remained unconvinced of the merit of Gandhi's forceful advocacy that everyone should spin at home with the "charka," the primitive spinning wheel. For Gandhi this practice was an important part of India's self-realization. "The spinning-wheel gradually became," as his biographer B. R. Nanda writes, "the center of rural uplift in the Gandhian scheme of Indian economics."[13] Tagore found the alleged economic rationale for this scheme quite unrealistic. As Romain Rolland noted, Rabindranath "never tires of criticizing the charka." In this economic judgment, Tagore was probably right. Except for the rather small specialized market for high-quality spun cloth, it is hard to make economic sense of hand-spinning, even with wheels less primitive than Gandhi's charka. Hand-spinning as a widespread activity can survive only with the help of heavy government subsidies.[14] However, Gandhi's advocacy of the charka was not based only on economics. He wanted everyone to spin for "thirty minutes every day as a sacrifice," seeing this as a way for people who are better off to identify themselves with the less fortunate. He was impatient with Tagore's refusal to grasp this point:

> *The poet lives for the morrow, and would have us do likewise .... "Why should I, who have no need to work for food, spin?" may be the question asked. Because I am eating what does not belong to me. I am living on the spoliation of my countrymen. Trace the source of every coin that finds its way into your pocket, and you will realise the truth of what I write. Every one must spin. Let Tagore spin like the others. Let him burn his foreign clothes; that is the duty today. God will take care of the morrow.*[15]

If Tagore had missed something in Gandhi's argument, so did Gandhi miss the point of Tagore's main criticism. It was not only that the charka made little economic sense, but also, Tagore thought, that it was not the way to make people reflect on anything: "The charka does not require anyone to think; one simply turns the wheel of the antiquated invention endlessly, using the minimum of judgment and stamina."

Celibacy and Personal Life

Tagore and Gandhi's attitudes toward personal life were also quite different. Gandhi was keen on the virtues of celibacy, theorized about it, and, after some years of conjugal life, made a private commitment — publicly announced — to refrain from sleeping with his wife. Rabindranath's own attitude on this subject was very different, but he was gentle about their disagreements:

> *[Gandhiji] condemns sexual life as inconsistent with the*
> *moral progress of man, and has a horror of sex as great*
> *as that of the author of The Kreutzer Sonata, but, unlike*
> *Tolstoy, he betrays no abhorrence of the sex that tempts his*
> *kind. In fact, his tenderness for women is one of the noblest*
> *and most consistent traits of his character, and he counts*
> *among the women of his country some of his best and truest*
> *comrades in the great movement he is leading.*

Tagore's personal life was, in many ways, an unhappy one. He married in 1883, lost his wife in 1902, and never remarried. He sought close companionship, which he did not always get (perhaps even during his married life — he wrote to his wife, Mrinalini: "If you and I could be comrades in all our work and in all our thoughts it would be splendid, but we cannot attain all that we desire"). He maintained a warm friendship with, and a strong Platonic attachment to, the literature-loving wife, Kadambari, of his elder brother, Jyotirindranath. He dedicated some poems to her before his marriage, and several books afterward, some after her death (she committed suicide, for reasons that are not fully understood, at the age of twenty-five, four months after Rabindranath's wedding). Much later in life, during his tour of Argentina in 1924–1925, Rabindranath came to know the talented and beautiful Victoria Ocampo, who later became the publisher of the literary magazine *Sur*. They became close friends, but it appears that Rabindranath deflected the possibility of a passionate relationship into a confined intellectual one.[16] His friend Leonard Elmhirst, who accompanied Rabindranath on his Argentine tour, wrote:

*Besides having a keen intellectual understanding of his
books, she was in love with him — but instead of being
content to build a friendship on the basis of intellect, she
was in a hurry to establish that kind of proprietary right
over him which he absolutely would not brook.*

Ocampo and Elmhirst, while remaining friendly, were both quite rude in what they wrote about each other. Ocampo's book on Tagore (of which a Bengali translation was made from the Spanish by the distinguished poet and critic Shankha Ghosh) is primarily concerned with Tagore's writings but also discusses the pleasures and difficulties of their relationship, giving quite a different account from Elmhirst's, and never suggesting any sort of proprietary intentions.

Victoria Ocampo, however, makes it clear that she very much wanted to get physically closer to Rabindranath: "Little by little he [Tagore] partially tamed the young animal, by turns wild and docile, who did not sleep, dog-like, on the floor outside his door, simply because it was not done."[17] Rabindranath, too, was clearly very much attracted to her. He called her "Vijaya" (the Sanskrit equivalent of Victoria), dedicated a book of poems to her, *Purabi* — an "evening melody," and expressed great admiration for her mind ("like a star that was distant"). In a letter to her he wrote, as if to explain his own reticence:

*When we were together, we mostly played with words
and tried to laugh away our best opportunities to see
each other clearly ... Whenever there is the least sign of
the nest becoming a jealous rival of the sky [,] my mind,
like a migrant bird, tries to take ... flight to a distant shore.*

Five years later, during Tagore's European tour in 1930, he sent her a cable: "Will you not come and see me." She did. But their relationship did not seem to go much beyond conversation, and their somewhat ambiguous correspondence continued over the years. Written in 1940, a year before his death at eighty, one of the poems in *Sesh Lekha* ("Last Writings"), seems to be about her: "How I wish I could once again find my way to that foreign land where waits for me the message of love!/... Her language I knew not, but what her eyes said will forever remain eloquent in its anguish."[18] However indecisive, or confused, or awkward Rabindranath may have been, he certainly did not share Mahatma Gandhi's censorious views of sex. In fact, when it came to social policy, he advocated contraception and family planning while Gandhi preferred abstinence.

Science and the People

Gandhi and Tagore severely clashed over their totally different attitudes toward science. In January 1934, Bihar was struck by a devastating earthquake, which killed thousands of people. Gandhi, who was then deeply involved in the fight against untouchability (the barbaric system inherited from India's divisive past, in which "lowly people" were kept at a physical distance), extracted a positive lesson from the tragic event. "A man like me," Gandhi argued, "cannot but believe this earthquake is a divine chastisement sent by God for our sins" — in particular the sins of untouchability. "For me there is a vital connection between the Bihar calamity and the untouchability campaign."

Tagore, who equally abhorred untouchability and had joined Gandhi in the movements against it, protested against this interpretation of an event that had caused suffering and death to so many innocent people, including children and babies. He also hated the epistemology implicit in seeing an earthquake as caused by ethical failure. "It is," he wrote, "all the more unfortunate because this kind of unscientific view of [natural] phenomena is too readily accepted by a large section of our countrymen."

The two remained deeply divided over their attitudes toward science. However, while Tagore believed that modern science was essential to the understanding of physical phenomena, his views on epistemology were interestingly heterodox. He did not take the simple "realist" position often associated with modern science. The report of his conversation with Einstein, published in *The New York Times* in 1930, shows how insistent Tagore was on interpreting truth through observation and reflective concepts. To assert that something is true or untrue in the absence of anyone to observe or perceive its truth, or to form a conception of what it is, appeared to Tagore to be deeply questionable. When Einstein remarked, "If there were no human beings any more, the Apollo Belvedere no longer would be beautiful?" Tagore simply replied, "No." Going further — and into much more interesting territory — Einstein said, "I agree with regard to this conception of beauty, but not with regard to truth." Tagore's response was: "Why not? Truth is realized through men."[19]

Tagore's epistemology, which he never pursued systematically, would seem to be searching for a line of reasoning that would later be elegantly developed by Hilary Putnam, who has argued: "Truth depends on conceptual schemes and it is nonetheless 'real truth.'"[20] Tagore himself said little to explain his convictions, but it is important to take account of his heterodoxy, not only because his speculations were invariably interesting, but also because they illustrate how his support for any position, including his strong interest in science, was accompanied by critical scrutiny.

Nationalism and Colonialism

Tagore was predictably hostile to communal sectarianism (such as a Hindu orthodoxy that was antagonistic to Islamic, Christian, or Sikh perspectives). But even nationalism seemed to him to be suspect. Isaiah Berlin summarizes well Tagore's complex position on Indian nationalism:

> *Tagore stood fast on the narrow causeway, and did not betray his vision of the difficult truth. He condemned romantic overattachment to the past, what he called the tying of India to the past "like a sacrificial goat tethered to a post," and he accused men who displayed it — they seemed to him reactionary — of not knowing what true political freedom was, pointing out that it is from English thinkers and English books that the very notion of political liberty was derived. But against cosmopolitanism he maintained that the English stood on their own feet, and so must Indians. In*

1917 he once more denounced the danger of 'leaving
everything to the unalterable will of the Master,' be he
brahmin or Englishman.[21]

The duality Berlin points to is well reflected also in Tagore's attitude toward cultural diversity. He wanted Indians to learn what is going on elsewhere, how others lived, what they valued, and so on, while remaining interested and involved in their own culture and heritage. Indeed, in his educational writings the need for synthesis is strongly stressed. It can also be found in his advice to Indian students abroad. In 1907 he wrote to his son-in-law Nagendranath Gangulee, who had gone to America to study agriculture:

> *To get on familiar terms with the local people is a part of*
> *your education. To know only agriculture is not enough;*
> *you must know America too. Of course if, in the process*
> *of knowing America, one begins to lose one's identity and*
> *falls into the trap of becoming an Americanised person*
> *contemptuous of everything Indian, it is preferable to stay*
> *in a locked room.*

Tagore was strongly involved in protest against the Raj on a number of occasions, most notably in the movement to resist the 1905 British proposal to split in two the province of Bengal, a plan that was eventually withdrawn following popular resistance. He was forthright in denouncing the brutality of British rule in India, never more so than after the Amritsar massacre of April 13, 1919, when 379 unarmed people at a peaceful meeting were gunned down by the army, and two thousand more were wounded. Between April 23 and 26, Rabindranath wrote five agitated letters to C. F. Andrews, who himself was extremely disturbed, especially after he was told by a British civil servant in India that thanks to this show of strength, the "moral prestige" of the Raj had "never been higher."

A month after the massacre, Tagore wrote to the Viceroy of India, asking to be relieved of the knighthood he had accepted four years earlier:

*The disproportionate severity of the punishments inflicted
upon the unfortunate people and the methods of carrying
them out, we are convinced, are without parallel in the
history of civilized governments, barring some
conspicuous exceptions, recent and remote. Considering
that such treatment has been meted out to a population,
disarmed and resourceless, by a power which has the
most terribly efficient organisation for destruction of
human lives, we must strongly assert that it can claim no
political expediency, far less moral justification … The
universal agony of indignation roused in the hearts of our
people has been ignored by our rulers — possibly
congratulating themselves for imparting what they imagine
as salutary lessons … I for my part want to stand, shorn
of all special distinctions, by the side of those of my
countrymen who for their so-called insignificance are
liable to suffer a degradation not fit for human beings.*

Both Gandhi and Nehru expressed their appreciation of the important part Tagore took in the national struggle. It is fitting that after independence, India chose a song of Tagore ("Jana Gana Mana Adhinayaka," which can be roughly translated as "the leader of people's minds") as its national anthem. Since Bangladesh would later choose another song of Tagore ("Amar Sonar Bangla") as its national anthem, he may be the only one ever to have authored the national anthems of two different countries.

Tagore's criticism of the British administration of India was consistently strong and grew more intense over the years. This point is often missed, since he made a special effort to dissociate his criticism of the Raj from any denigration of British — or Western — people and culture. Mahatma Gandhi's well-known quip in reply to a question, asked in England, on what he thought of Western civilization ("It would be a good idea") could not have come from Tagore's lips. He would understand the provocations to which Gandhi was responding — involving cultural conceit as well as imperial tyranny.

D. H. Lawrence supplied a fine example of the former: "I become more and more surprised to see how far higher, in reality, our European civilization stands than the East, Indian and Persian, ever dreamed of ... This fraud of looking up to them — this wretched worship-of-Tagore attitude is disgusting." But, unlike Gandhi, Tagore could not, even in jest, be dismissive of Western civilization.

Even in his powerful indictment of British rule in India in 1941, in a lecture which he gave on his last birthday, and which was later published as a pamphlet under the title *Crisis in Civilization*, he strains hard to maintain the distinction between opposing Western imperialism and rejecting Western civilization. While he saw India as having been "smothered under the dead weight of British administration" (adding "another great and ancient civilization for whose recent tragic history the British cannot disclaim responsibility is China"), Tagore recalls what India has gained from "discussions centred upon Shakespeare's drama and Byron's poetry and above all ... the large-hearted liberalism of nineteenth-century English politics." The tragedy, as Tagore saw it, came from the fact that what "was truly best in their own civilization, the upholding of dignity of human relationships, has no place in the British administration of this country." "If in its place they have established, baton in hand, a reign of 'law and order,' or in other words a policeman's rule, such a mockery of civilization can claim no respect from us."

Critique of Patriotism

Rabindranath rebelled against the strongly nationalist form that the independence movement often took, and this made him refrain from taking a particularly active part in contemporary politics. He wanted to assert India's right to be independent without denying the importance of what India could learn — freely and profitably — from abroad. He was afraid that a rejection of the West in favor of an indigenous Indian tradition was not only limiting in itself; it could easily turn into hostility to other influences from abroad, including Christianity, which came to parts of India by the fourth century;

Judaism, which came through Jewish immigration shortly after the fall of Jerusalem, as did Zoroastrianism through Parsi immigration later on (mainly in the eighth century), and, of course — and most importantly — Islam, which has had a very strong presence in India since the tenth century.

Tagore's criticism of patriotism is a persistent theme in his writings. As early as 1908, he put his position succinctly in a letter replying to the criticism of Abala Bose, the wife of a great Indian scientist, Jagadish Chandra Bose: "Patriotism cannot be our final spiritual shelter; my refuge is humanity. I will not buy glass for the price of diamonds, and I will never allow patriotism to triumph over humanity as long as I live." His novel *Ghare Baire* (The Home and the World) has much to say about this theme. In the novel, Nikhil, who is keen on social reform, including women's liberation, but cool toward nationalism, gradually loses the esteem of his spirited wife, Bimala, because of his failure to be enthusiastic about anti-British agitations, which she sees as a lack of patriotic commitment. Bimala becomes fascinated with Nikhil's nationalist friend Sandip, who speaks brilliantly and acts with patriotic militancy, and she falls in love with him. Nikhil refuses to change his views: "I am willing to serve my country; but my worship I reserve for Right which is far greater than my country. To worship my country as a god is to bring a curse upon it."[22]

As the story unfolds, Sandip becomes angry with some of his countrymen for their failure to join the struggle as readily as he thinks they should ("Some Mohamedan traders are still obdurate"). He arranges to deal with the recalcitrants by burning their meager trading stocks and physically attacking them. Bimala has to acknowledge the connection between Sandip's rousing nationalistic sentiments and his sectarian — and ultimately violent-actions. The dramatic events that follow (Nikhil attempts to help the victims, risking his life) include the end of Bimala's political romance.

This is a difficult subject, and Satyajit Ray's beautiful film of *The Home and the World* brilliantly brings out the novel's tensions, along with the human affections and disaffections of the story. Not surprisingly, the story has had many detractors, not just among dedicated nationalists in India. Georg Lukács found Tagore's novel to be "a petit bourgeois yarn of the shoddiest kind," "at the intellectual service of the British police," and "a contemptible caricature of Gandhi." It would, of course, be absurd to think of Sandip as Gandhi, but the novel gives a "strong and gentle" warning, as Bertolt Brecht noted in his diary, of the corruptibility of nationalism, since it is not even-handed. Hatred of one group can lead to hatred of others, no matter how far such feeling may be from the minds of large-hearted nationalist leaders like Mahatma Gandhi.

Admiration and Criticism of Japan

Tagore's reaction to nationalism in Japan is particularly telling. As in the case of India, he saw the need to build the self-confidence of a defeated and humiliated people, of people left behind by developments elsewhere, as was the case in Japan before its emergence during the nineteenth century. At the beginning of one of his lectures in Japan in 1916 ("Nationalism in Japan"), he observed that "the worst form of bondage is the bondage of dejection, which keeps men hopelessly chained in loss of faith in themselves." Tagore shared the admiration for Japan widespread in Asia for demonstrating the ability of an Asian nation to rival the West in industrial development and economic progress. He noted with great satisfaction that Japan had "in giant strides left centuries of inaction behind, overtaking the present time in its foremost achievement." For other nations outside the West, he said, Japan "has broken the spell under which we lay in torpor for ages, taking it to be the normal condition of certain races living in certain geographical limits."

But then Tagore went on to criticize the rise of a strong nationalism in Japan, and its emergence as an imperialist nation. Tagore's outspoken criticisms did not please Japanese audiences and, as E. P. Thompson wrote, "the welcome given to him on his first arrival soon cooled."[23] Twenty-two years later, in 1937, during the Japanese war on China, Tagore received a letter from Rash Behari Bose, an anti-British Indian revolutionary then living in Japan, who sought Tagore's approval for his efforts there on behalf of Indian independence, in which he had the support of the Japanese government. Tagore replied:

> *Your cable has caused me many restless hours, for it hurts me very much to have to ignore your appeal. I wish you had asked for my cooperation in a cause against which my spirit did not protest. I know, in making this appeal, you counted on my great regard for the Japanese for I, along with the rest of Asia, did once admire and look up to Japan and did once fondly hope that in Japan Asia had at last discovered its challenge to the West, that Japan's new strength would be consecrated in safeguarding the culture of the East against alien interests. But Japan has not taken long to betray that rising hope and repudiate all that seemed significant in her wonderful, and, to us symbolic, awakening, and has now become itself a worse menace to the defenceless peoples of the East.*

How to view Japan's position in the Second World War was a divisive issue in India. After the war, when Japanese political leaders were tried for war crimes, the sole dissenting voice among the judges came from the Indian judge, Radhabinod Pal, a distinguished jurist. Pal dissented on various grounds, among them that no fair trial was possible in view of the asymmetry of power between the victor and the defeated. Ambivalent feelings in India toward the Japanese military aggression, given the unacceptable nature of British imperialism, possibly had a part in predisposing Pal to consider a perspective different from that of the other judges.

More tellingly, Subhas Chandra Bose (no relation of Rash Behari Bose), a leading nationalist, made his way to Japan during the war via Italy and Germany after escaping from a British prison; he helped the Japanese to form units of Indian soldiers, who had earlier surrendered to the advancing Japanese army, to fight on the Japanese side as the "Indian National Army." Rabindranath had formerly entertained great admiration for Subhas Bose as a dedicated nonsectarian fighter for

Indian independence.[24] But their ways would have parted when Bose's political activities took this turn, although Tagore was dead by the time Bose reached Japan.

Tagore saw Japanese militarism as illustrating the way nationalism can mislead even a nation of great achievement and promise. In 1938 Yone Noguchi, the distinguished poet and friend of Tagore (as well as of Yeats and Pound), wrote to Tagore, pleading with him to change his mind about Japan. Rabindranath's reply, written on September 12, 1938, was altogether uncompromising:

> *It seems to me that it is futile for either of us to try to convince the other, since your faith in the infallible right of Japan to bully other Asiatic nations into line with your Government's policy is not shared by me .... Believe me, it is sorrow and shame, not anger, that prompt me to write to you. I suffer intensely not only because the reports of Chinese suffering batter against my heart, but because I can no longer point out with pride the example of a great Japan.*

He would have been much happier with the postwar emergence of Japan as a peaceful power. Then, too, since he was not free of egotism, he would also have been pleased by the attention paid to his ideas by the novelist Yasunari Kawabata and others.[25]

International Concerns

Tagore was not invariably well-informed about international politics. He allowed himself to be entertained by Mussolini in a short visit to Italy in May–June 1926, a visit arranged by Carlo Formichi, professor of Sanskrit at the University of Rome. When he asked to meet Benedetto Croce, Formichi said, "Impossible! Impossible!" Mussolini told him that Croce was "not in Rome." When Tagore said he would go "wherever he is," Mussolini assured him that Croce's whereabouts were unknown.

Such incidents, as well as warnings from Romain Rolland and other friends, should have ended Tagore's flirtation with Mussolini more quickly than it did. But only after he received graphic accounts of the brutality of Italian fascism from two exiles, Gaetano Salvemini and Gaetano Salvadori, and learned more of what was happening in Italy, did he publicly denounce the regime, publishing a letter to the *Manchester Guardian* in August. The next month, *Popolo d'Italia*, the magazine edited by Benito Mussolini's brother, replied: "Who cares? Italy laughs at Tagore and those who brought this unctuous and insupportable fellow in our midst."

With his high expectations of Britain, Tagore continued to be surprised by what he took to be a lack of official sympathy for international victims of aggression. He returned to this theme in the lecture he gave on his last birthday, in 1941:

> *While Japan was quietly devouring North China, her act of wanton aggression was ignored as a minor incident by the veterans of British diplomacy. We have also witnessed from this distance how actively the British statesmen acquiesced in the destruction of the Spanish Republic.*

But distinguishing between the British government and the British people, Rabindranath went on to note "with admiration how a band of valiant Englishmen laid down their lives for Spain."

Tagore's view of the Soviet Union has been a subject of much discussion. He was widely read in Russia. In 1917 several Russian translations of *Gitanjali* (one edited by Ivan Bunin, later the first Russian Nobel Laureate in Literature) were available, and by the late 1920s many of the English versions of his work had been rendered into Russian by several distinguished translators. Russian versions of his work continued to appear: Boris Pasternak translated him in the 1950s and 1960s.

When Tagore visited Russia in 1930, he was much impressed by its development efforts and by what he saw as a real commitment to eliminate poverty and economic inequality. But what impressed him most was the expansion of basic education across the old Russian empire. In *Letters from Russia*, written in Bengali and published in 1931, he unfavorably compares the acceptance of widespread illiteracy in India by the British administration with Russian efforts to expand education:

> In stepping on the soil of Russia, the first thing that caught
> my eye was that in education, at any rate, the peasant
> and the working classes have made such enormous
> progress in these few years that nothing comparable has
> happened even to our highest classes in the course of
> the last hundred and fifty years ... The people here are
> not at all afraid of giving complete education even to
> Turcomans of distant Asia; on the contrary, they are
> utterly in earnest about it. [26]

When parts of the book were translated into English in 1934, the under-secretary for India stated in Parliament that it was "calculated by distortion of the facts to bring the British Administration in India into contempt and disrepute," and the book was then promptly banned. The English version would not be published until after independence.

Education and Freedom

The British Indian administrators were not, however, alone in trying to suppress Tagore's reflections on Russia. They were joined by Soviet officials. In an interview with *Izvestia* in 1930, Tagore sharply criticized the lack of freedom that he observed in Russia:

> I must ask you: Are you doing your ideal a service by
> arousing in the minds of those under your training anger,
> class-hatred, and revengefulness against those whom
> you consider to be your enemies? ... Freedom of mind is
> needed for the reception of truth; terror hopelessly kills
> it.... For the sake of humanity I hope you may never
> create a vicious force of violence, which will go on
> weaving an interminable chain of violence and cruelty ...
> You have tried to destroy many of the other evils of [the
> czarist] period. Why not try to destroy this one also?

The interview was not published in *Izvestia* until 1988 — nearly sixty years later.[27] Tagore's reaction to the Russia of 1930 arose from two of his strongest commitments: his uncompromising belief in the importance of "freedom of mind" (the source of his criticism of the Soviet Union), and his conviction that the expansion of basic education is central to social progress (the source of his praise, particularly in contrast to British-run India). He identified the lack of basic education as the fundamental cause of many of India's social and economic afflictions:

> *In my view the imposing tower of misery which today*
> *rests on the heart of India has its sole foundation in the*
> *absence of education. Caste divisions, religious conflicts,*
> *aversion to work, precarious economic conditions — all*
> *centre on this single factor.*

It was on education (and on the reflection, dialogue, and communication that are associated with it), rather than on, say, spinning "as a sacrifice" ("the charka does not require anyone to think"), that the future of India would depend.

Tagore was concerned not only that there be wider opportunities for education across the country (especially in rural areas where schools were few), but also that the schools themselves be more lively and enjoyable. He himself had dropped out of school early, largely out of boredom, and had never bothered to earn a diploma. He wrote extensively on how schools should be made more attractive to boys and girls and thus more productive. His own co-educational school at Santiniketan had many progressive features. The emphasis here was on self-motivation rather than on discipline, and on fostering intellectual curiosity rather than competitive excellence.

Much of Rabindranath's life was spent in developing the school at Santiniketan. The school never had much money, since the fees were very low. His lecture honoraria, "$700 a scold," went to support it, as well as most of his Nobel Prize money. The school received no support from the government, but did get help from private citizens — even Mahatma Gandhi raised money for it.

The dispute with Mahatma Gandhi on the Bihar earthquake touched on a subject that was very important to Tagore: the need for education in science as well as in literature and the humanities. At Santiniketan, there were strong "local" elements in its emphasis on Indian traditions, including the classics, and in the use of Bengali rather than English as the language of instruction. At the same time there were courses on a great variety of cultures, and study programs devoted to China, Japan, and the Middle East. Many foreigners came to Santiniketan to study or teach, and the fusion of studies seemed to work.

I am partial to seeing Tagore as an educator, having myself been educated at Santiniketan. The school was unusual in many different ways, such as the oddity that classes, excepting those requiring a laboratory, were held outdoors (whenever the weather permitted). No matter what we thought of Rabindranath's belief that one gains from being in a natural setting while learning (some of us argued about this theory), we typically found the experience of outdoor schooling extremely attractive and pleasant. Academically, our school was not particularly exacting (often we did not have any examinations at all), and it could not, by the usual academic standards, compete with some of the better schools in Calcutta. But there was something remarkable about the ease with which class discussions could move from Indian traditional literature to contemporary as well as classical Western thought, and then to the culture of China or Japan or elsewhere. The school's celebration of variety was also in sharp contrast with the cultural conservatism and separatism that has tended to grip India from time to time. The cultural give and take of Tagore's vision of the contemporary world has close parallels with the vision of Satyajit Ray, also an alumnus of Santiniketan who made several films based on Tagore's stories.[28] Ray's words about Santiniketan in 1991 would have greatly pleased Rabindranath:

> *I consider the three years I spent in Santiniketan as the most fruitful of my life ... Santiniketan opened my eyes for the first time to the splendours of Indian and Far Eastern art. Until then I was completely under the sway of Western art, music and literature. Santiniketan made me the combined product of East and West that I am.*[29]

India Today

At the fiftieth anniversary of Indian independence, the reckoning of what India had or had not achieved in this half century was a subject of considerable interest: "What has been the story of those first fifty years?" (as Shashi Tharoor asked in his balanced, informative, and highly readable account of India: *From Midnight to the Millennium*).[30] If Tagore were to see the India of today, more than half a century after independence, nothing perhaps would shock him so much as the continued illiteracy of the masses. He would see this as a total betrayal of what the nationalist leaders had promised during the struggle for independence — a promise that had figured even in Nehru's rousing speech on the eve of independence in August 1947 (on India's "tryst with destiny").

In view of his interest in childhood education, Tagore would not be consoled by the extraordinary expansion of university education, in which India sends to its universities six times as many people per unit of population as does China. Rather, he would be stunned that, in contrast to East and Southeast Asia, including China, half the adult population and two-thirds of Indian women remain unable to read or write. Statistically reliable surveys indicate that even in the late 1980s, nearly half of the rural girls between the ages of twelve and fourteen did not attend any school for a single day of their lives.[31]

This state of affairs is the result of the continuation of British imperial neglect of mass education, which has been reinforced by India's traditional elitism, as well as upper-class-dominated contemporary politics (except in parts of India such as Kerala, where anti-upper-caste movements have tended to concentrate on education as a great leveller). Tagore would see illiteracy and the neglect of education not only as the main source of India's continued social backwardness, but also as a great constraint that restricts the possibility and reach of economic development in India (as his writings on rural development forcefully make clear). Tagore would also have strongly felt the need for a greater commitment — and a greater sense of urgency — in removing endemic poverty.

At the same time, Tagore would undoubtedly find some satisfaction in the survival of democracy in India, in its relatively free press, and in general from the "freedom of mind" that post-independence Indian politics has, on the whole, managed to maintain. He would also be pleased by the fact noted by the historian E. P. Thompson (whose father Edward Thompson had written one of the first major biographies of Tagore):[32]

> *All the convergent influences of the world run through*
> *this society: Hindu, Moslem, Christian, secular; Stalinist,*
> *liberal, Maoist, democratic socialist, Gandhian. There is*
> *not a thought that is being thought in the West or East that*
> *is not active in some Indian mind.*[33]

Tagore would have been happy also to see that the one governmental attempt to dispense generally with basic liberties and political and civil rights in India, in the 1970s, when Prime Minister Indira Gandhi (ironically, herself a former student at Santiniketan) declared an "emergency," was overwhelmingly rejected by the Indian voters, leading to the precipitate fall of her government.

Rabindranath would also see that the changes in policy that have eliminated famine since independence had much to do with the freedom to be heard in a democratic India. In Tagore's play *Raja O Rani* ("The King and the Queen"), the sympathetic Queen eventually rebels against the callousness of state policy toward the hungry. She begins by inquiring about the ugly sounds outside the palace, only to be told that the noise is coming from "the coarse, clamorous crowd who howl unashamedly for food and disturb the sweet peace of the palace." The Viceregal office in India could have taken a similarly callous view of Indian famines, right up to the easily preventable Bengal famine of 1943, just before independence, which killed between two and three million people. But a government in a multi-party democracy, with elections and free newspapers, cannot any longer dismiss the noise from "the coarse, clamorous crowd."[34]

Unlike Gandhi, Rabindranath would not resent the development of modern industries in India, or the acceleration of technical progress, since he did not want India to be shackled to the turning of "the wheel of an antiquated invention." Tagore was concerned that people not be dominated by machines, but he was not opposed to making good use of modern technology. "The mastery over the machine," he wrote in *Crisis in Civilization*, "by which the British have consolidated their sovereignty over their vast empire, has been kept a sealed book, to which due access has been denied to this helpless country." Rabindranath had a deep interest in the environment — he was particularly concerned about deforestation and initiated a "festival of tree-planting" (*vriksha-ropana*) as early as 1928. He would want increased private and government commitments to environmentalism; but he would not derive from this position a general case against modern industry and technology.

On Cultural Separation

Rabindranath would be shocked by the growth of cultural separatism in India, as elsewhere. The "openness" that he valued so much is certainly under great strain right now — in many countries. Religious fundamentalism still has a relatively small following in India; but various factions seem to be doing their best to increase their numbers. Certainly religious sectarianism has had much success in some parts of India (particularly in the west and the north). Tagore would see the expansion of religious sectarianism as being closely associated with an artificially separatist view of culture.

He would have strongly resisted defining India in specifically Hindu terms, rather than as a "confluence" of many cultures. Even after the partition of 1947, India is still the third-largest Muslim country in the world, with more Muslims than in Bangladesh, and nearly as many as in Pakistan. Only Indonesia has substantially more followers of Islam. Indeed, by pointing to the immense heterogeneousness of India's cultural background and its richly diverse history, Tagore had argued that the "idea of India" itself militated against a culturally separatist view — "against the intense consciousness of the separateness of one's own people from others."

Tagore would also oppose the cultural nationalism that has recently been gaining some ground in India, along with an exaggerated fear of the influence of the West. He was uncompromising in his belief that human beings could absorb quite different cultures in constructive ways:

> *Whatever we understand and enjoy in human products*
> *instantly becomes ours, wherever they might have their*
> *origin. I am proud of my humanity when I can*
> *acknowledge the poets and artists of other countries as*
> *my own. Let me feel with unalloyed gladness that all the*
> *great glories of man are mine. Therefore it hurts me*
> *deeply when the cry of rejection rings loud against the*
> *West in my country with the clamour that Western*
> *education can only injure us.*

In this context, it is important to emphasize that Rabindranath was not short of pride in India's own heritage, and often spoke about it. He lectured at Oxford, with evident satisfaction, on the importance of India's religious ideas — quoting both from ancient texts and from popular poetry (such as the verses of the sixteenth-century Muslim poet Kabir). In 1940, when he was given an honorary doctorate by Oxford University, in a ceremony arranged at his own educational establishment in Santiniketan ("In Gangem Defluit Isis," Oxford helpfully explained), to the predictable "volley of Latin" Tagore responded "by a volley of Sanskrit," as Marjorie Sykes, a Quaker friend of Rabindranath, reports. Her cheerful summary of the match, "India held its own," was not out of line with Tagore's pride in Indian culture. His welcoming attitude to Western civilization was reinforced by this confidence: he did not see India's culture as fragile and in need of "protection" from Western influence.

In India, he wrote, "circumstances almost compel us to learn English, and this lucky accident has given us the opportunity of access into the richest of all poetical literatures of the world." There seems to me much force in Rabindranath's argument for clearly distinguishing between the injustice of a serious asymmetry of power (colonialism being a prime example of this) and the importance nevertheless of appraising Western culture in an open-minded way, in colonial and postcolonial territories, in order to see what uses could be made of it.

Rabindranath insisted on open debate on every issue, and distrusted conclusions based on a mechanical formula, no matter how attractive that formula might seem in isolation (such as "This was forced on us by our colonial masters — we must reject it," "This is our tradition — we must follow it," "We have promised to do this — we must fulfill that promise," and so on). The question he persistently asks is whether we have reason enough to want what is being proposed, taking everything into account. Important as history is, reasoning has to go beyond the past. It is in the sovereignty of reasoning — fearless reasoning in freedom — that we can find Rabindranath Tagore's lasting voice.[35]

Endnotes

1. Rabindranath Tagore, *The Religion of Man* (London: Unwin, 1931, 2nd edition, 1961), p. 105. The extensive interactions between Hindu and Muslim parts of Indian culture (in religious beliefs, civic codes, painting, sculpture, literature, music, and astronomy) have been discussed by Kshiti Mohan Sen in *Bharate Hindu Mushalmaner Jukto Sadhana* (in Bengali) (Calcutta: Visva-Bharati, 1949, extended edition, 1990) and *Hinduism* (Penguin, 1960).

2. Rabindranath's father Debendranath had in fact joined the reformist religious group, the Brahmo Samaj, which rejected many contemporary Hindu practices as aberrations from the ancient Hindu texts.

3. *Selected Letters of Rabindranath Tagore,* edited by Krishna Dutta and Andrew Robinson (Cambridge University Press, 1997). This essay draws on my Foreword to this collection. For important background material on Rabindranath Tagore and his reception in the West, see also the editors' *Rabindranath Tagore: The Myriad-Minded Man* (St. Martin's Press, 1995), and *Rabindranath Tagore: An Anthology* (Picador, 1997).

4. See *Romain Rolland and Gandhi Correspondence*, with a Foreword by Jawaharlal Nehru (New Delhi: Government of India, 1976), pp. 12–13.

5. On Dartington Hall, the school, and the Elmhirsts, see Michael Young, *The Elmhirsts of Dartington: The Creation of an Utopian Community* (Routledge, 1982).

6. I have tried to analyze these "exotic" approaches to India (along with other Western approaches) in "India and the West," *The New Republic*, June 7, 1993, and in "Indian Traditions and the Western Imagination," *Daedalus*, Spring 1997.

7. Yasunari Kawabata, *The Existence and Discovery of Beauty*, translated by V. H. Viglielmo (Tokyo: The Mainichi Newspapers, 1969), pp. 56–57.

8. W. B. Yeats, "Introduction," in Rabindranath Tagore, *Gitanjali* (London: Macmillan, 1913).

9. Tagore himself vacillated over the years about the merits of his own translations. He told his friend Sir William Rothenstein, the artist: "I am sure you remember with what reluctant hesitation I gave up to your hand my manuscript of Gitanjali, feeling sure that my English was of that amorphous kind for whose syntax a school-boy could be reprimanded." These — and related — issues are discussed by Nabaneeta Dev Sen, "The 'Foreign Reincarnation' of Rabindranath Tagore," *Journal of Asian Studies*, 25 (1966), reprinted, along with other relevant papers, in her *Counterpoints: Essays in Comparative Literature* (Calcutta: Prajna, 1985).

10. The importance of ambiguity and incomplete description in Tagore's poetry provides some insight into the striking thesis of William Radice (one of the major English translators of Tagore) that "his blend of poetry and prose is all the more truthful for being incomplete" ("Introduction" to his *Rabindranath Tagore: Selected Short Stories*, Penguin, 1991, p. 28).

11. Satyajit Ray, the film director, has argued that even in Tagore's paintings, "the mood evoked ... is one of a joyous freedom" (Ray, "Foreword," in Andrew Robinson, *The Art of Rabindranath Tagore*, London: André Deutsch, 1989).

12. Reported in Amita Sen, *Anando Sharbokaje* (in Bengali) (Calcutta: Tagore Research Institute, 2nd edition, 1996), p. 132.

13. B. R. Nanda, *Mahatma Gandhi* (Oxford University Press, 1958; paperback, 1989), p. 149.

14. The economic issues are discussed in my *Choice of Techniques* (Blackwell, 1960), Appendix D.

15. Mohandas Gandhi, quoted by Krishna Kripalani, *Tagore: A Life* (New Delhi: Orient Longman, 1961, 2nd edition, 1971), pp. 171–172.

16. For fuller accounts of the events, see Dutta and Robinson, *Rabindranath Tagore: The Myriad-Minded Man*, Chapter 25, and Ketaki Kushari Dyson, *In Your Blossoming Flower-Garden: Rabindranath Tagore and Victoria Ocampo* (New Delhi: Sahitya Akademi, 1988).

17. Published in English translation in *Rabindranath Tagore: A Centenary Volume, 1861–1961* (New Delhi: Sahitya Akademi, 1961), with an Introduction by Jawaharlal Nehru.

18. English translation from Krishna Kripalani, *Tagore: A Life*, p. 185.

19. "Einstein and Tagore Plumb the Truth," *The New York Times Magazine*, August 10, 1930; republished in Dutta and Robinson, *Selected Letters of Rabindranath Tagore*.

20. Hilary Putnam, *The Many Faces of Realism* (Open Court, 1987). On related issues, see also Thomas Nagel, *The View from Nowhere* (Oxford University Press, 1986).

21. Isaiah Berlin, "Rabindranath Tagore and the Consciousness of Nationality," *The Sense of Reality: Studies in Ideas and Their History* (New York: Farrar, Straus and Giroux, 1997), p. 265.

22. Martha Nussbaum initiates her wide-ranging critique of patriotism (in a debate that is joined by many others) by quoting this passage from *The Home and the World* (in Martha C. Nussbaum et al., *For Love of Country*, edited by Joshua Cohen, Beacon Press, 1996, pp. 3–4).

23. E. P. Thompson, Introduction, to *Tagore's Nationalism* (London, Macmillan, 1991), p. 10.

24. For a lucid and informative analysis of the role of Subhas Chandra Bose and his brother Sarat in Indian politics, see Leonard A. Gordon, *Brothers against the Raj: A Biography of Indian Nationalists Sarat and Subhas Chandra Bose* (Columbia University Press, 1990).

25. Kawabata made considerable use of Tagore's ideas, and even built on Tagore's thesis that it "is easier for a stranger to know what it is in [Japan] which is truly valuable for all mankind" (*The Existence and Discovery of Beauty*, pp. 55–58).

26. Tagore, *Letters from Russia*, translated from Bengali by Sasadhar Sinha (Calcutta: Visva-Bharati, 1960), p. 108.

27. It was, however, published in the *Manchester Guardian* shortly after it was meant to be published in the *Izvestia*. On this, see: Dutta and Robinson, *Rabindranath Tagore: The Myriad-Minded Man*, p. 297.

28. Satyajit Ray, *Our Films, Their Films* (Calcutta: Disha Book/Orient Longman, third edition, 1993). I have tried to discuss these issues in my Satyajit Ray Memorial Lecture, "Our Culture, Their Culture," *The New Republic*, April 1, 1996.

29. *The Guardian*, August 1, 1991.

30. Arcade Publishing, 1997, p. 1.

31. On this and related issues, see Jean Drèze and Amartya Sen, *India: Economic Development and Social Opportunity* (Clarendon Press/Oxford University Press, 1996), particularly Chapter 6, and also Drèze and Sen, editors, *Indian Development: Selected Regional Perspectives* (Clarendon Press/ Oxford University Press, 1996).

32. Edward Thompson, *Rabindranath Tagore: Poet and Dramatist* (Oxford University Press, 1926).

33. Quoted in Shashi Tharoor, *India*, p. 9.

34. I have tried to discuss the linkage between democracy, political incentives, and prevention of disasters in *Resources, Values and Development*

(Harvard University Press, 1984, reprinted 1997), Chapter 19, and in my presidential address to the American Economic Association, "Rationality and Social Choice," *American Economic Review*, 85 (1995).

35. For helpful discussions I am most grateful to Akeel Bilgrami, Sissela Bok [Harvard Professor; the daughter of Gunnar Myrdal, recipient of The Bank of Sweden Prize in Economic Sciences in Memory of Alfred Nobel 1974, and Alva Myrdal, who was awarded The Nobel Peace Price in 1982], Sugata Bose, Supratik Bose, Krishna Dutta, Rounaq Jahan, Salim Jahan, Marufi Khan, Andrew Robinson, Nandana Sen, Gayatri Chakravorty Spivak, and Shashi Tharoor.

✳

Tagore at Santiniketan in 1923.

Ehrensvärd Photo. Courtesy of the Royal Library collections, Stockholm, and the Nobel Foundation.

Nelson MANDELA
Photo courtesy of the Nobel Foundation.

Nelson Mandela and the Rainbow of Culture

❄

Anders Hallengren

Equality and Pluralism

After 27 years in prison, Nelson Mandela negotiated the dismantling of the apartheid regime in South Africa, settled an agreement on universal suffrage and democratic elections, and became the first black president of the country in 1994. When he entered into office, he was aware of the universal importance of this success, but he was also humbled by the focus on his person as a symbol of international and historical dimensions. After all, during the years 1952–1990, he had made only three public appearances, and numerous people of different nations had contributed to the cause. Indeed, Africa had been liberated from colonialism during his prison years. The truth of the ancient Bantu adage *umuntu ngumuntu ngabantu* (which he translates as "we are people through other people") often came to his mind. And he saw, perhaps clearer than most of his contemporaries, the inevitability of "mutual interdependence" in the human condition, that "the common ground is greater and more enduring than the differences that divide." The background of the development of this vision is remarkable and diverse. From his African heritage, from his country's turbulent history, from his own formal education in "colonial" schools, and from his vicissitudes in the confines of Robben Island and other prisons, Mandela emerged a man with a singular vision.

The Development of "Colour-blindness"

Starting his fight for liberation of the blacks as an aggressive young African pugilist and nationalist in the early 1940s, Mandela had not always deemed that democratic progress must rest on equality, pluralism, and multiethnicity. What made him later stand out from other South African leaders, and made him finally emerge victorious, was precisely his vision of a state that belongs equally to all its different peoples, nations, and tribes, whether Afrikaan, English, or Zulu. Being himself a leader belonging to the Xhosa-speaking people, he eventually transcended the idea of national liberty, and he attracted Indians, Jews, and other segments of the multicoloured population to the cause. Countering the racist suppression of the blacks, he avoided, unlike many other revolutionaries of the continent, acceding to a basically or exclusively black or tribal liberation movement.

This vision, sometimes referred to as "colour-blindness," was partly inspired by Marxists, drawing on European ideologies and influenced from abroad. They were internationalists, not nationalists, and fought for a class, not for a race. The South African Communist Party, originally founded in 1920, and the kernel of which was a small group of Jewish immigrants and English nonconformists, was to influence the African National Congress (ANC) in such a direction, without ever succeeding in turning Nelson Mandela, Oliver Tambo, and other leaders into members of their party. This influence was theoretical and ideological, based on reading, hearsay, and revolutionary tradition.

Through the years, it also became increasingly economic, in terms of financial support from parts of the communist, the socialist, and the social-democratic world — including Sweden, Norway, and India as well as the USSR. Mandela's party, the ANC, was completely pragmatic in its views of material means, however. It accepted succour and aid wherever it came from, whether from Libya (as it happens, one of

Mandela's grandchildren was baptized "Gadaffi"), Iraq, diamond investors, or multinational corporations, sticking to its cause and never deviating from its course. The Sotho maxim "many rills make a big river" often was in Mandela's mind. As a matter of fact, and quite contrary to contemporary European and American views, Nelson Mandela and the ANC remained ideologically independent while their financial dependence grew. Nevertheless, as a result of this focus and political imbalance, the ANC became a pawn in great-power politics, which delayed Mandela's release and peaceful reform in South Africa until the winding-up of the Eastern bloc in 1989–90.

The Legacy of Mahatma Gandhi and Pandit Nehru

Another source of transnational perspectives and ideas of coexistence in harmony was the likewise oppressed Indian population of South Africa, many of whom traced their origin from the indentured labourers shipped to local sugar cane fields by the British colonial authorities. The confidence and faith of contemporary Indian freedom fighters rested upon the belief in the *Satyagraha* (truth-power) preached and put into effect by Mahatma Gandhi, a vision that had freed India in 1947. The vision of Gandhi was kept alive and thrived in South Africa, where Gandhi himself had lived and worked for many years (1893–1914). Nelson Mandela's early encounters with these more peaceful Hindu, as well as Moslem, activists and their ideologies of emancipation seriously complicated his view of African liberation, and a close bond between the ANC and South Africa's Indian population developed over time. This personal encounter with other people's liberation movements in South Africa, eventually — almost of necessity, as it were — made the ANC leadership turn multicultural and multireligious, bound together by a common goal and based on that "common ground" Mandela often refers to.

When the present writer attended a function where Nelson Mandela received The Gandhi/King Award for Non-violence in 1999, the prize was presented by Ms. Ela Gandhi, granddaughter of Mahatma Gandhi and member of the South African Parliament. She described Mandela as "the man who completed the anti-colonial movement began by Mahatma Gandhi" and, indeed, "the living legacy of Mahatma Gandhi; the Gandhi of South Africa". Furthermore, India on 16 March 2001 conferred The International Gandhi Peace Prize on Nelson Mandela at a grand ceremony at the presidential palace in New Delhi. "In honouring Mandela," President Shri K. R. Narayanan said, "we are paying tribute to an unusual hero in the Gandhian mould, who personifies the triumph of the human spirit over forces of oppression." In his acceptance speech, Mandela recalled India's support to South Africa during the years of struggle against apartheid.

Penetrating even deeper into historical memory, Mohandas Karamchand Gandhi's speech in Johannesburg in 1908 makes us aware of the persistent dream that propelled a progressive movement and is as yet to come true on a universal basis. On that occasion, Gandhi for the first time expressed his dream of a free South Africa, a vision that was to echo throughout the century:

> *"If we look into the future [of South Africa], is it not*
> *a heritage we have to leave to posterity, that all the*
> *different races commingle and produce a civilisation*
> *that perhaps the world has not yet seen?"*

This speech is quoted in *Gandhi and South Africa 1914–1948*, a collection of articles edited by E. S. Reddy and Gopalkrishna Gandhi, published in India with the dedication "To Nelson Mandela and His Colleagues". In that book, Gandhiji is in fact mentioned as "South Africa's Gift to India." M. K. Gandhi, Attorney in South Africa, was in a deep sense succeeded by Nelson Mandela, Johannesburg Attorney. Or, as Nelson Mandela said in September 1992, when he had been released from prison and democratic reform was on the agenda:

> *"Gandhiji was a South African and his memory deserves*
> *to be cherished now and in post-apartheid South Africa.*
> *We must never lose sight of the fact that the Gandhian*
> *philosophy may be a key to human survival in the*
> *twenty-first century."*

Manilal Gandhi, the son of Mahatma, remained in his father's house in Natal, South Africa, and the Indian reform movement encouraged the chaotic but awakening *civil disobedience* campaigns shaking large parts of South Africa in the early 1950s, in particular the Defiance Campaign of 1952, where the internal passport laws and other apartheid measures were challenged. These originally Indian-inspired peaceful resistance initiatives were met by violence, and the movements gradually turned underground, where they eventually grew. Finally the ANC reacted by building a military command, the Umkhonto we Sizwe ("Spear of the Nation"), preparing itself for guerrilla war. Contrary to Gandhi, and unlike the ANC leader Albert Lutuli, who received the Nobel Peace Prize in 1960, or Archbishop Desmund Tutu, Peace Prize Laureate of 1984, Nelson Mandela long felt forced to advocate unavoidable revolutionary violence, considering it as a counter-violence or a justified uprising against iniquitous laws. Mandela eventually dressed himself in camouflage uniform and was in combat training abroad (in Ethiopia), as were many other revolutionaries. Banned by the South African regime from attending public gatherings from late 1952 on, he nevertheless had efficiently and ingeniously continued his work as an organiser through undercover actions, escaping the police in ever-changing disguises, one of the favourites being that of a chauffeur. The years before his final imprisonment in 1962, he was nicknamed "The Black Pimpernel."

Still, the Indian heritage of peaceful resistance was present in Mandela's mind, and one of his closest fellow-combatants and conspirators was the Indian Ahmed Kathrada, with whom he was to spend a quarter of a century in jail. Among Mandela's models and teachers, however, Jawaharlal Nehru — more militant than Gandhi and a politician of a practical turn — had become the prominent figure.

We should recall that his Indian friends evoked Mandela's interest in India at the time when that state was in the process of being liberated from British colonialism. Nehru, who in 1947 became the first Prime Minister of India, had for two decades urged the Indians in South Africa to join forces with the black Africans, and India was the first country to introduce sanctions against the apartheid regime. One of Mandela's most famous phrases and titles, "No Easy [Long] Walk to Freedom", was a quote from Nehru, whose hardships and determination he identified with. While in jail, Mandela was greatly encouraged by receiving the Indian Nehru Prize of 1979, and in his speech of thanks — in his absence read in New Delhi by the exiled Oliver Tambo — he emphasized his indebtedness to Nehru. Their fates were similar, too. When later on speaking of his release from prison, Nelson Mandela has said: "I was helped when preparing for my release by the biography of Pandit Nehru, who wrote of what happens when you leave jail. My daughter Zinzi says that she grew up without a father, who, when he returned, became a father of the nation."

The Ancient African Source of Wisdom

The deeper basis for the development of Mandela's universal vision and his confidence was his faith in education. More than everything else, Mandela emphasized education as important for his understanding, as well as for the growth of his people and the development of humanity in general. Furthermore, and consequently, he championed the availability of information and learning to all people. Part of his own peculiar schooling for life was, he points out, his long imprisonment as well as the ancient traditions of the Thembu people and his early years in missionary schools. If we venture to approach "the Making of Mandela," the script that develops is his lifelong education, in a deep and singular way combining the core of African and European traditions.

Born in 1918 in the little village of Mvezo in Qunu in southern Transkei, as the son of the chief Gadla Henry Mphakanyiswa, Rolihlahla Mandela was the great-grandson of Ngubengcuka, the glorious king of the Thembu people who had died in 1832 before British occupation of the territory. Mandela's clan, the Madiba, considered itself as dethroned royalty. The upbringing of the son was accordingly, and when his father died, the Thembu ruler Jongintaba and his wife No-England took care of the boy's education.

A central concept in this Xhosa-speaking culture, as in Bantu tradition in general, is *Ubuntu*, fraternity. This implies compassion and open-mindedness and is opposed to individualism and egotism. Nelson learned how the belligerent and ruthless foreign occupants had destroyed the ancient peace and harmony that reigned among the different Xhosa tribes in the past. The effect was his persistent longing for the reestablishment of precolonial concord, the renaissance of the golden age of Africa. Accordingly, Mandela was always to prefer his clan name, Madiba, as his personal name. When he helped form the ANC Youth League in the Bantu Men's Social Centre in Johannesburg 1944, the manifesto underscored that the African, in contrast to the white man, regards the universe as an organic whole in progress towards harmony, where individual parts exist only as aspects of this universal unity. There is continuity in this emphasis and in this African heritage, since the essence of the concept ubuntu was to recur and be expanded on in the new South African Constitution of 1996.

On the other hand, his mother Nosekeni Fanny had become a devout Christian and sent him to a missionary primary school (where he was given the English name Nelson). As a result, he was introduced and initiated into African teachings and traditions by his regal mentor, while at the same time he was the first ever in his family to attend the local Methodist school. His education had become pronouncedly Afro-European.

Learning the Western Canon

The Methodist schools not only inspired the temperance and discipline of Mandela's lifestyle but proved to be a training ground for liberation, even more so when he moved on to the Clarkebury School, which had been founded in 1825 under the rule of his ancestor king Ngubengcuka. Without these schools for Africans, there would have been neither transfer of power nor any black presidency, Mandela has pointed out.

He advanced to the all-British Healdtown High School, where the principal was a descendant of Lord Wellington, and then to the South

African Native College at Fort Hare. There he encountered teachers who made an impression for life: Prof. Z. K. Matthews, the first black pupil to graduate from the college and a relative of Mandela; and the headmaster Alexander Kerr, who drilled the students in English literature. Matthews was to produce the ANC Freedom Charter. Kerr made the pupils see the relevancy of classical European literature in a contemporary African context and got Mandela and his classmates to quote English poets wherever they went. Mandela was forever to praise his teachers, and these school experiences are part and parcel of the stress on education in the ANC Program of Action of 1949, which, in its basic aims, established the educational means and the cultural ends of liberation:

— To unite the cultural with the educational and national struggle;
— To establish a national academy of arts and sciences.

After Healdtown followed studies in law and social sciences at Witwaterstand University in 1943–49. These were never officially finished but were taken up again when, as with some other prisoners, he was later allowed to study through correspondence at the University of South Africa and, in 1980, at London University. In fact, the real school for life started when he was arrested in 1962, and indeed this is literally true, since he spent the next thirty years studying and teaching.

Robben Island University

Robben Island became a campus for political prisoners. They had to work outdoors in an isolated lime quarry, where, when left to themselves in mine shafts, they discussed their different views and taught each other what they knew, year after year. Mandela wanted the spirit of a university to reign, and regular lectures were arranged in secrecy. Thus, the prison would be called The Robben Island University or, even later, Nelson Mandela University. From debates with prisoners and white warders through the years, Mandela grew to a widened ideological awareness from which he would draw when arguing with the government on a new South African constitution. In particular, he studied Afrikaans and learnt to understand the mind-set of the Boer minority through discussions with warders and staff. Mandela identified with these descendants of Dutch seventeenth-century immigrants and saw that he himself, under other circumstances, could have taken views similar to theirs. And he always appreciated their fight against the English in the Boer War, which seemed partly to parallel and herald his own fight against oppression. From this understanding, Mandela's remarkable spirit of reconciliation was drawn, when Africans and Afrikaners finally formed a coalition government.

Mandela saw historical connections in all present events, and he knew that the British had imprisoned Xhosa chiefs at Robben Island already in the early nineteenth century. These ancestors and predecessors, first and foremost among them the ruler Makanna, whose warriors had almost defeated the English at Grahamstown in 1819, made him feel the long and strong line behind him, pressing and urging him onwards.

As times went by, the prisoners found more and more ways of communicating with each other, and the interchange of ideas and knowledge grew. They formed a motley company, where the variety proved to be a great asset to them all. There was the famous South African poet Dennis Brutus, a constant source of inspiration. There were the most foul-mouthed felons as well as an intellectual elite. As important as the presence of African culture, however, was the society of dead English poets.

Shakespeare the Revolutionary

If Africa was the source of their illumination, European literature served as a propelling dynamo for their nerve. Of particular pertinence was William Shakespeare, who was constantly quoted and discussed. Were they not wronged and deceived by a criminal ruler as had been Hamlet, the prince of Denmark? The subjects of a treacherous Macbeth, and like Macbeth was the South African regime not doomed to destruction? Or, were they not rightfully conspiring against a despotic Julius Caesar? Did not the future rest upon them? Did they really have the guts to shoulder their responsibility? Already an ANC Youth League flyer of 1944 had concluded in the summoning words of Julius Caesar:

The fault ... is not in our stars,
But in ourselves, that we are underlings.

The Bible and Shakespeare: the missionary schools had been effective. The Pan African Congress President Robert Sobukwe had translated Shakespeare's *Macbeth* into Zulu, just as Tanzania's liberator Julius Nyerere had translated *Julius Caesar* (for whom he was named) into Swahili. Shakespeare was an active part in the political play. When Harold Macmillan, the conservative premier of England, had delivered a speech in Capetown in 1960, a newspaper cartoon pictured him afterwards with a caption picked from *Julius Caesar*:

O, pardon me, thou bleeding piece of earth,
That I am meek and gentle with these butchers!

Mandela, who always forgave but never forgot, was to refer to this political cartoon when he spoke in London in 1996. In jail, Mandela's extraordinary memory was a mine, and to him, Shakespeare was The Writer. While in custody, a subject of debate could be whether George Bernard Shaw could measure up to Shakespeare. When Mandela was waiting for the final verdict after the extended legal proceedings, and was prepared to be sentenced to death, Shakespeare's counsel in *Measure for Measure* was a source of courage:

Be absolute for death; either death or life
Shall thereby be the sweeter.

Unlike the Bible, The Qur'an, the Bhagavadgita, or *Das Kapital* by Karl Marx, Shakespeare was a common denominator for the prisoners at Robben Island. Only a few of them were Christian believers; a few were Moslems or Hindus; a few, communists; and their origins were different. They all knew Shakespeare, however. Furthermore, from the point of view of the penitentiary authorities, four-hundred-year-old Elizabethan dramas were not considered dangerous reading. But

Shakespeare was profoundly political and always had something of acute importance to say. Militant passages drawn from *Coriolanus* or *Henry V* excited them, and *Julius Caesar* stirred them to rebellion. They knew their destiny, and Mandela's favourite passage read:

> *Cowards die many times before their deaths;*
> *The valiant never taste of death but once.*

Free to study classical drama, the prisoners at Robben Island staged a more than two-thousand-year-old Greek tragedy, the *Antigone* of Sophocles, in which earthly power is challenged with reference to a higher law. In that production under lock and key, Mandela played the part of Creon, the tyrant.

In fact, Mandela began thinking of political organization in terms of dramaturgy. In late June 1976, he noted: "It is often desirable for one not to describe events, but to put the reader in the atmosphere in which the whole drama was played out right inside the theatre, so that he can see with his own eyes the actual stage, all the actors and their costumes, follow their movements, listen to what they say and sing, and to study the facial expressions and the spontaneous reaction of the audience as the drama unfolds." Thabo Mbeki, Ahmed Kathrada, Walter Sisulu, Neville Alexander, and all the others were inspired by words that clearly expressed their thoughts and notions, and at once these became tools in a revolutionary process and cutting rejoinders on a political scene.

From Henley to Heaney

Another source of encouragement was the words of a Victorian English poet, William Ernest Henley (1849–1903). Decade after decade, the unforgettable lines of the poem "Invictus", "unconquerable," were on Mandela's lips:

> *OUT of the night that covers me,*
> *Black as the Pit from pole to pole,*
> *I thank whatever gods may be*
> *For my unconquerable soul.*

In the fell clutch of circumstance
I have not winced nor cried aloud.
Under the bludgeonings of chance
My head is bloody, but unbowed.

Beyond this place of wrath and tears
Looms but the Horror of the shade,
And yet the menace of the years
Finds, and shall find, me unafraid.

It matters not how strait the gate,
How charged with punishments the scroll,
I am the master of my fate:
I am the captain of my soul.

At Robben Island, Mandela recited this poem and taught other prisoners these defiant lines; reading such words "puts life in you", Mandela says. From the British philosopher Bertrand Russell, who had been jailed for his protests against nuclear weapons, Mandela had drawn the arguments of defiance, when conscience and civil laws do not agree. From the Russian novelist and idealist Lev Tolstoy, Kathrada and Mandela got similar support. At times, they also identified with the endless waiting of the protagonists in Samuel Beckett's play *Waiting for Godot*.

This literary heritage came with an awareness of the blackness derived from the same source — an often-discussed paradox: they were fighting foreign rule with foreign maxims. Among Xhosa poets, Mandela's favourite was Krune Mkwayi (Mqhayi), whose haunting lament to the Prince of Wales in 1925 has been translated into English by Robert Mshengu Kavanagh:

You sent us the truth, denied us the truth;
You sent us the life, deprived us of life;
You sent us the light, we sit in the dark,
Shivering, benighted in the bright noonday sun.

They were also heartened by the existence of contemporary South African authors like Nadine Gordimer, a devoted supporter. But when they were released and a new government was formed, in which prisoners were transformed into ministers of state, they still quoted English poetry, such as the lines of the contemporary Nobel Laureate Seamus Heaney:

> *The longed-for tidal wave*
> *of justice can rise up,*
> *And hope and history rhyme*

The Liberation of Faith

In December 1999, the Parliament of the World's Religions assembled for the third time since 1893, this time in Capetown. Mandela, who was scheduled to go to the United States, changed his itinerary, anxious to attend. He considered the event of utmost importance. He also had something he wanted to say. In his concluding speech, delivered on December 5, he pointed out the importance of religion to the liberation of his country and his own indebtedness, countering the propaganda that had for half a century pictured him as a godless infidel:

> *"In our country, my generation is the product of religious*
> *education. We grew up at a time when the government of*
> *this country owed its duty only to whites, a minority of less*
> *than fifteen percent. They took no interest whatsoever in our*
> *education. It was religious institutions, whether Christian,*
> *Moslem, Hindu, or Jewish, in the context of our country;*
> *they are the people who bought land, who built schools, who*
> *employed teachers and paid them. Without church, without*
> *religious institutions, I would never have been here today.*
> *It was for that reason, that when I was ready to go to the*
> *United States on the first of this month, an engagement*
> *which had been arranged for quite some time, when my*
> *comrade, Ibrahim, told me about this occasion, I said: I will*
> *change my own itinerary, so I will have the opportunity of*
> *appearing here. But I must also add, that I appreciate the*
> *importance of religion. Apart from the background I have*

given you, you have to have been in a South African jail
under apartheid, where you can see a cruelty of human
beings to others in a naked form. But it was again religious
institutions, Hindus, Moslems, leaders of the Jewish faith,
Christians, it was them who gave us the hope that one day,
we would come out, we would return. And in prison, the
religious institutions raised funds for our children, who were
arrested in thousands and thrown into jail, and many of
them one day left prison at a high level of education, because
of this support we got from religious institutions. And that
is why we so respect religious institutions. And we try as
much as we can to read the literature, which outlines the
fundamental principles of human behaviour. And, like the
Bhagadgita, the Qur'an, the Bible, and other important
religious documents. And I say this, so you should
understand, that the propaganda that has been made for
example about the liberation movement is completely untrue.
Because religion was one of the motivating factors in
everything that we did."

Mandela also looked onwards. Concerning the prospects of the twenty-first century, he pointed out: "No less than in any other period of history, religion will have a crucial role to play in guiding and inspiring humanity to meet the enormous challenges that we face."

The Rainbow Government

In conclusion, the outlook and horizon of Mandela and his sympathisers, who were to form the ideological centre of the new administration of 1994, was stamped by East and West, by European and Asian, as well as genuine African influences. This eclecticism cohered and interacted with the cultural variety of their country and their different origins, which propelled their joint endeavours. The final outcome was the Rainbow Nation, the most multiethnic government ever formed.

Successfully curbing and harnessing the *indlovu ayipatha*, the dangerous elephant in the shape of the apartheid regime, the rebels finally reached their goal and settled the dispute between Africans and Afrikaners. In achieving this, Mandela agitated against black domination. He did not argue for a turning back to a glorious African society of bygone times but called for a completely new kind of state, a multiethnic democracy without match, constituted by a manifold of cultures having equal rights.

Thus, the ANC leadership, as reorganized when Mandela was released in 1990 and could officially take on command, consisted of a cross section of races, including seven Indians, seven "Coloureds," and seven whites. Likewise, and in harmony with this, a broad cultural and political basis marked the government of 1994. Ministers of state were blacks, whites, Indians, Coloureds, Muslims, Christians, communists, liberals, conservatives. Apart from three Indian Muslims, there were two Hindus in Mandela's government. Never had such a cabinet been seen in Africa or elsewhere. Many prominent posts were occupied by neither Africans nor by Afrikaners: there were Dullah Omar, Minister of Justice; the Minister of Water Kader Asmal; the Minister of Finance Trevor Manuel ... all widening the arch of the rainbow government.

This was the realization of a political program outlined almost half a century earlier. The date can be determined with some certainty. In "The Programme of Action: Statement of Policy Adopted at the ANC Annual Conference 17 December 1949," there is hardly a trace of the future multicultural view. In The Freedom Charter, however, adopted at the Congress of the People, Kliptown, on 26 June 1955, the ideology is emerging:

> *"The rights of the people shall be the same, regardless of race, colour or sex ... All National Groups Shall have Equal Rights! ... Men and women of all races shall receive equal pay for equal work;"*

The idea is explicit but not yet fully developed in Mandela's speech for the defence from the dock at the Rivonia Trial in 1963, where he empasized, "I have fought against white domination and I have fought against black domination. I have cherished the ideal of a democratic and free society in which all persons live together in harmony and with equal opportunities." In his speech in Harlem, New York, in June 1990, the recently discharged prisoner repeated these words, adding, "Death to racism! Glory to the sisterhood and brotherhood of peoples throughout the world!" The basic concept of *ubuntu* had by then been expanded beyond all limits of nation, race, faith and gender.

The deed and document of this ideology is the 1996 Constitution of the Republic of South Africa, one of the most modern and radical in the world regarding human rights. It includes an extensive Bill of Rights (Ch. 2) where the numerous paragraphs on equality stand out, for example:

"The state may not unfairly discriminate directly or indirectly against anyone on one or more grounds, including race, gender, sex, pregnancy, marital status, ethnic or social origin, colour, sexual orientation, age, disability, religion, conscience, belief, culture, language and birth." In Ch. 1, "Founding Provisions," the official languages are enumerated; all are permitted for national government and provincial government use: "The official languages of the Republic are Sepedi, Sesotho, Setswana, siSwati, Tshivenda, Xitsonga, Afrikaans, English, isiNdebele, isiXhosa and isiZulu."

It also states that the president determines the national anthem of the Republic by proclamation. Such proclamation had already occurred. Today, *Nkosi Sikelel' iAfrika*, the old song of the revolutionaries, sang by Mandela and his comrades decade after decade, echoes throughout the republic in its chorus of equal tongues. The first line concludes the Preamble of the Constitution:

> *Nkosi Sikelel' iAfrika*
> *Morena boloka setjhaba sa heso*
> *God se'n Suid-Afrika*
> *Mudzimu fhatutshedza Afurika*
> . *Hosi katekisa Afrika*
> *God bless South Africa.*

Educational Goals

The Constitution, which is also supplemented by an Action Plan, stresses the importance of education, mentioning this already in the Bill of Rights:

> *"Everyone has the right —*
> *a. to a basic education, including adult basic education; and*
> *b. to further education, which the state, through reasonable measures, must make progressively available and accessible."*

Again, the Constitution of 1996 can be seen as the fulfilment of the ANC Freedom Charter of 1955:

> *"All people shall have equal right to use their own languages, and to develop their own folk culture and customs ... The aim of education shall be to teach the youth to love their people and their culture, to honour human brotherhood, liberty and peace; Education shall be free, compulsory, universal and equal for all children; Higher education and technical training shall be opened to all by means of state allowances and scholarships awarded on the basis of merit; Adult illiteracy shall be ended by a mass state education plan."*

In this spirit, people today may enter The Nelson Mandela Gateway to Robben Island Museum Heritage Services (its museum building completed in December 2001) and go by boat to the penitentiary which Mandela turned into a university. At Robben Island, we also find the model Robben Island Primary School, a school of integration and international awareness.

What's In a Name?

Mandela never wanted places and things to be named after him. The few exceptions are schools and educational institutions, such as the Nelson Mandela School of Medicine in Durban and Dr. Nelson Mandela High School in the Western Cape; or, to look abroad,

'Nelson Mandela School' in the German Democratic Republic, founded in 1984, and Nelson Mandela School of Public Policy and Urban Affairs in Baton Rouge, LA, USA.

A significant and very telling feature, however, witnessing his care for the future generations, is the Nelson Mandela Children's Fund, established when President Nelson Mandela pledged one-third of his salary for five years.

Conclusion

Thus, we see how one man's remarkable life has reached its fulfillment and has blossomed into a national vision. Inspired by myriad influences, taking the best from both his Native heritage, from the example of foreign freedom movements, and even from the history and literature of his oppressors, Nelson Mandela forged a vision of humanity that encompasses all peoples and that sets the hallmark for the rest of the world.

When the Norwegian Nobel Committee decided to award the Nobel Peace Prize for 1993 to Nelson R. Mandela and Frederik Willem de Klerk, it was pointed out that their achievement was made by "looking ahead to South African reconciliation instead of back at the deep wounds of the past." The committee also observed that South Africa has been the very symbol of racially conditioned suppression, and that the peaceful termination of the apartheid regime accordingly "points the way to the resolution of similar deep-rooted conflicts elsewhere in the world."

In his Nobel lecture, Nelson Mandela referred to the organic world-view expressed already in the manifesto of 1944, calling himself a mere representative of the millions of people across the globe who "recognised that an injury to one is an injury to all;" which is the essence of *ubuntu* philosophy universally applied.

❄

Bibliography

✳

V. S. Naipaul

WORKS

The Mystic Masseur. London: André Deutsch, 1957

Miguel Street. London: Deutsch, 1959

A House for Mr. Biswas. London: Deutsch, 1961

The Middle Passage: Impressions of Five Societies — British, French and Dutch in the West Indies and South America. London: Deutsch, 1962

Mr. Stone and the Knights Companion. London: Deutsch, 1963

A Flag on the Island. London: Deutsch, 1967

The Mimic Men. London: Deutsch, 1967

The Loss of El Dorado: A History. London: Deutsch, 1969

In a Free State. London: Deutsch, 1971

The Overcrowded Barracoon and Other Articles. London: Deutsch, 1972

Guerrillas. London: Deutsch, 1975

India: A Wounded Civilization. London: Deutsch, 1977

A Bend in the River. London: Deutsch, 1979

A Congo Diary. Los Angeles: Sylvester & Orphanos, 1980

Among the Believers: An Islamic Journey. London: Deutsch, 1981

The Enigma of Arrival. London: Viking, 1987

India: A Million Mutinies Now. London: Heinemann, 1990

A Way in the World. London: Heinemann, 1994

Beyond Belief: Islamic Excursions among the Converted Peoples. London: Little, Brown, 1998

Reading and Writing: A Personal Account. New York: The New York Review of Books, [London: Bloomsbury] distributor, 2000

Half a Life. London: Picador, 2001

The Writer and the World: Essays. Edited by Pankaj Mishra. London: Picador, 2002

Literary Occasions: Essays. Introduced and edited by Pankaj Mishra. London: Picador, 2003

Magic Seeds. London: Picador, 2004

SECONDARY LITERATURE

Ashcroft, Bill, Gareth Griffiths, Helen Tiffin (eds.). *The Empire Writes Back: Theory and Practice in Post-colonial Literatures,* 2nd ed. London: Routledge, 2002 [1989]

Barnouw, Dagmar. *Naipaul's Strangers.* Bloomington: Indiana Univ. Press, 2003

Dissanayake, Wimal. *Self and Colonial Desire: Travel Writings of V. S. Naipaul.* New York: P. Lang, 1993

Feder, Lillian. *Naipaul's Truth: The Making of a Writer.* Lanham, MD: Rowman & Littlefield, 2001

Gorra, Michael. *After Empire: Scott, Naipaul, Rushdie.* Chicago, IL: Univ. of Chicago Press, 1997

Hamner, Robert D. (ed.). *Critical Perspectives on V. S. Naipaul.* London: Heinemann, 1979

Hassan, Dolly Zulakha. *V. S. Naipaul and the West Indies.* New York: P. Lang, 1989

Hayward, Helen. *The Enigma of V. S. Naipaul: Sources and Contexts.* New York: Palgrave Macmillan, 2002

Jarvis, Kelvin. *V. S. Naipaul: A Selective Bibliography with Annotations, 1957–1987.* Metuchen, NJ: Scarecrow, 1989

Jussawalla, Feroza (ed.). *Conversations with V. S. Naipaul.* Jackson: Univ. Press of Mississippi, 1997

Kelly, Richard. *V. S. Naipaul*. New York: Continuum, 1989

Khan, Akhtar Jamal. *V. S. Naipaul: A Critical Study*. New Delhi: Creative Books, 1998

King, Bruce. *V. S. Naipaul*. Basingstoke: Macmillan, 1993

Levy, Judith. *V. S. Naipaul: Displacement and Autobiography*. New York: Garland, 1995

Nasta, Susheila (ed.). *Writing Across Worlds: Contemporary Writers Talk*. London: Routledge, 2004

Nightingale, Peggy. *Journey through Darkness: The Writing of V. S. Naipaul*. St. Lucia: Univ. of Queensland Press, 1987

Panwar, Purabi. *India in the Works of Kipling, Forster and Naipaul: Postcolonial Revaluations*. New Delhi: Pencraft International, 2000

Singh, Manjit Inder. *The Poetics of Alienation and Identity: V. S. Naipaul and George Lamming*. Delhi: Ajanta, 1998

Theroux, Paul. *Sir Vidia's Shadow: A Friendship across Five Continents*. Boston: Houghton Mifflin, 1998

Weiss, Timothy F. *On the Margins: The Art of Exile in V. S. Naipaul*. Amherst: Univ. of Massachusetts Press, 1992

White, Landeg. *V. S. Naipaul*. New York: Barnes & Noble, 1975

Nadine Gordimer

Novels

The Lying Days. London: Gollancz; New York: Simon, 1953

A World of Strangers. London: Gollancz, 1958

Occasion for Loving. New York: Viking; London: Gollancz, 1963

The Late Bourgeois World. New York: Viking; London: Gollancz, 1966

A Guest of Honour. New York: Viking, 1970

The Conservationist. London: Cape, 1974

Burger's Daughter. New York: Viking; London: Jonathan Cape, 1980

July's People. London: Jonathan Cape, 1981

A Sport of Nature. New York: Knopf; London: Jonathan Cape; Cape Town: D. Philip, 1987

My Son's Story. London: Bloomsbury; New York: Farrar, Strauss and Giroux, 1990

None to Accompany Me. London: Bloomsbury; New York: Farrar, Straus and Giroux, 1994

The House Gun. London: Bloomsbury; New York: Farrar, Straus and Giroux, 1998

The Pickup. London: Bloomsbury; New York: Farrar, Straus and Giroux, 2001

Short Story Collections

Face to Face. Johannesburg: Silver Leaf Books, 1949

The Soft Voice of the Serpent. New York: Simon & Schuster, 1952 (overlapping with *Face to Face*)

Six Feet of the Country. London: Gollancz, 1956

Friday's Footprint. London: Gollancz, 1960

Not for Publication. London: Gollancz, 1965

Livingstone's Companions. New York: Viking, 1971

Selected Stories. London: Jonathan Cape, 1975

A Soldier's Embrace. London: Jonathan Cape, 1980

Something Out There. London: Jonathan Cape, 1984

Crimes of Conscience. Oxford: Heineman, 1991

Jump and Other Stories. Cape Town: D. Philip; London: Bloomsbury, 1991

Why Haven't You Written; Selected Stories 1950–1972. London: Penguin, 1992

Loot and Other Stories. London: Bloomsbury; New York: Farrar, Straus and Giroux, 2003

SECONDARY LITERATURE

Bazin, Nancy Topping, and Marilyn Dallman Seymour (eds.). *Conversations with Nadine Gordimer*. Jackson: Univ. Press of Mississippi, 1990

Brahimi, Denise (ed.). *Nadine Gordimer: la femme, la politique et le roman*; préface de Claude Wauthier. Paris: Karthala; Johannesburg: IFAS, 2000

Clingman, Stephen. *The Novels of Nadine Gordimer: History from the Inside*. London: Bloomsbury, 1993

Driver, Dorothy (ed.). *Nadine Gordimer: A Bibliography of Primary and Secondary Sources, 1937–1992*. London: H. Zell, 1994

Ettin, Andrew V. (ed.). *Betrayals of the Body Politic: The Literary Commitments of Nadine Gordimer*. Charlottesville: Univ. Press of Virginia, 1993

Head, Dominic. *Nadine Gordimer*. Cambridge: Cambridge Univ. Press, 1994

King, Bruce (ed.). *The Later Fiction of Nadine Gordimer*. London: Macmillan, 1993

Newman, Judie (ed.). *Nadine Gordimer's Burger's Daughter: A Casebook*. Oxford: Oxford Univ. Press, 2003

Oliphant, Andries Walter (ed.). *A Writing Life: Celebrating Nadine Gordimer*. Photographs by David Goldblatt. London: Viking, 1998

Pettersson, Rose. *Nadine Gordimer's One Story of a State Apart*. Studia Anglistica Upsaliensia, 88, Uppsala University, Stockholm: Almqvist & Wiksell, 1995

Uledi Kamanga and Brighton J. *Cracks in the Wall: Nadine Gordimer's Fiction and the Irony of Apartheid*. Trenton, NJ: Africa World Press, 2002

Uraizee, Joya F. *This is No Place for a Woman: Nadine Gordimer, Nayantara Sahgal, Buchi Emecheta, and the Politics of Gender*. Trenton, NJ: Africa World Press, 1999

Wagner, Kathrin M. *Rereading Nadine Gordimer*. Bloomington: Indiana Univ. Press, 1994

Derek Walcott

Poetry

25 Poems. Port-of-Spain: Guardian Commercial Printery, 1948

Epitaph for the Young, XII Cantos. Bridgetown: Barbados Advocate, 1949

Poems. Kingston, Jamaica: City Printery, 1951

In a Green Night: Poems 1948–60. London: Jonathan Cape, 1962

Selected Poems. New York: Farrar, Straus and Giroux, 1964

The Castaway and Other Poems. London: Jonathan Cape, 1965

The Gulf and Other Poems. London: Jonathan Cape, 1969

Another Life. New York: Farrar, Straus and Giroux; London: Jonathan Cape, 1973

Sea Grapes. London: Cape; New York: Farrar, Straus and Giroux, 1976

The Star-Apple Kingdom. New York: Farrar, Straus and Giroux, 1979

Selected Poetry. Edited by Wayne Brown. London: Heinemann, 1981

The Fortunate Traveller. New York: Farrar, Straus and Giroux, 1981

The Caribbean Poetry of Derek Walcott, and the Art of Romare Bearden. New York: Limited Editions Club, 1983

Midsummer. New York: Farrar, Straus and Giroux, 1984

Collected Poems 1948–1984. New York: Farrar, Straus and Giroux, 1986

The Arkansas Testament. New York: Farrar, Straus and Giroux, 1987; London: Faber, 1988

Omeros. New York: Farrar, Straus and Giroux, 1990

The Bounty. London: Faber; New York: Farrar, Straus and Giroux, 1997

Tiepolo's Hound. New York: Farrar, Straus and Giroux, 2000

Drama

Harry Dernier. Bridgetown: Barbados Advocate, 1952

Dream on Monkey Mountain and Other Plays. New York: Farrar, Straus and Giroux, 1970

The Joker of Seville & O Babylon!. New York: Farrar, Straus and Giroux, 1978

Remembrance & Pantomine: Two Plays. New York: Farrar, Straus and Giroux, 1980

Three Plays. New York: Farrar, Straus and Giroux, 1986

The Odyssey. London: Faber; New York: Farrar, Straus and Giroux, 1993

The Haitian Trilogy. New York: Farrar, Straus and Giroux, 2002

Walker and the Ghost Dance. New York: Farrar, Straus and Giroux, 2002

Essays

What the Twilight Says. New York: Farrar, Straus and Giroux, 1998

SECONDARY LITERATURE

Breslin, Paul. *Nobody's Nation: Reading Derek Walcott.* Chicago: Univ. of Chicago Press, 2001

Brown, Stewart (ed.). *The Art of Derek Walcott.* Bridgend: Seren Books; Chester Springs, PA: Dufour, 1991

Burnett, Paula. *Derek Walcott: Politics and Poetics.* Univ. Press of Florida: Gainesville, FL, 2001

King, Bruce. *Derek Walcott: A Caribbean Life.* Oxford: Oxford Univ. Press, 2000

King, Bruce. *Derek Walcott & West Indian Drama: "Not Only a Playwright but a Company" — the Trinidad Theatre Workshop, 1959– 1993.* Oxford: Clarendon Press, 1995

Thieme, John. *Derek Walcott.* Manchester: Manchester Univ. Press, 1999

Walcott, Derek and William Baer. *Conversations with Derek Walcott.* Jackson: Univ. Press of Mississippi, 1996

Naguib Mahfouz

WORKS

[*'Abath Al-Aqdar,* 1939]. *La malédiction de Râ.* French transl. José M. Ruiz-Funes and Ahmed Mostefaï. Paris: l'Archipel, 1998

[*Radubis,* 1943]. *Rhadopis, La courtesana.* Spanish transl. María Luisa Prieto and Muhammmad al-Madkuri. Barcelona: Planeta-DeAgostini, 1998

[*Kifah Tiba,* 1944]. *La battaglia di Tebe.* Italian transl. by Anna Pagnini. Rome: Newton & Compton Editori, 2001

[*Al-Qahira Al-Gadida* (New Cairo), 1946]. *Nieuw Cairo.* Dutch transl. Djûke Poppinga. Breda: Uitgeverij De Geus, 1998

[*Khan Al-Khalili*, 1946]. *Le Cortège des vivants: Khan al-Khalili*. French transl. Faïza and Gilles Ladkany. Paris: Éditions Sindbad/ Actes Sud, 1999

[*Zuqaq Al-Midaqq*, 1947]. Translated as *Midaq Alley* by Trevor Le Gassick. Beirut: Khayat, 1966 / London: Heineman; Washington, D.C.: Three Continents Press, 1977

[*Al-Sarab* (Mirage), 1948]. *Chimères*. French transl. France Douvier Meyer. Paris: Denoël, 1992

[*Bidaya wa Nihaya*, 1949]. Translated as *The Beginning and the End* by Ramses Awad. ed. Mason Rossiter Smith. Cairo: The American Univ. in Cairo Press [AUC Press], 1985 / New York: Doubleday, 1989

[*Bayn Al-Qasrayn*, 1956]. Translated as *Palace Walk*. The Cairo Trilogy Part 1 by William Maynard Hutchins and Olive E. Kenny. Cairo: AUC Press, 1989 / New York: Doubleday, 1990

[*Qasr Al-Shawq*, 1957] *Palace of Desire*. The Cairo Trilogy Part 2 transl. by William Maynard Hutchins, Lorne M. Kenny and Olive E. Kenny. Cairo: AUC Press, 1991 / New York: Doubleday, 1991

[*Al-Sukkariyya*, 1957] *Sugar Street*. The Cairo Trilogy Part 3 transl. by William Maynard Hutchins and Angele Botros Samaan. Cairo: AUC Press, 1992 / New York: Doubleday, 1992. For a complete edition of the trilogy, see: *The Cairo Trilogy* [*al-Thulathiyya*], Cairo: AUC Press, 2001

[*Awlad Haratina* (Children of Our Quarter), 1959]. The original text was serialised in *Al-Ahram* in 1959, and printed in book form abroad — Beirut: Dar Al-Adab, 1967. Translated as *Children of Gebelawi* by Philip Stewart. London: Heinemann, 1981, and by Peter Theroux as *Children of the Alley*, New York: Doubleday 1996 / paperback ed: Cairo: AUC Press, 2001

[*Al-Liss wa-l-Kilab*, 1961]. *The Thief and the Dogs*. Transl. Trevor Le Gassick and M. M. Badawi. Cairo: AUC Press, 1984 / New York: Doubleday, 1989

[*Al-Summan wa-l-Kharif*, 1962]. Transl. *Autumn Quail* by Roger Allen, rev. John Rodenbeck. Cairo: AUC Press, 1985 / New York: Doubleday, 1990

[*Al-Tariq*, 1964]. *The Search*. Transl. Mohamed Islam, ed. Magdi Wahba. Cairo: AUC Press, 1987 / New York: Doubleday, 1991

[*Al-Shahhadh*, 1965]. Translated as *The Beggar* by Kristin Walker Henry and Nariman Khales Naili Al-Warraki. Cairo: AUC Press, 1986 / New York: Doubleday, 1990

[*Tharthara Fawq Al-Nil*, 1966]. Transl. *Adrift on the Nile* by Frances Liardet. Cairo: AUC Press, 1993 / London: Doubleday, 1993

[*Miramar*, 1967]. *Miramar*. Transl. Fatma Moussa-Mahmoud, edited and revised by Maged el Kommos and John Rodenbeck. London: Heinemann, 1978 / Cairo: AUC Press, 1978, 1993 / Washington, D.C.: Three Continents Press, 1983, 1992

[*Al-Maraya*, 1972]. *Mirrors*. Translated by Roger Allen. Minneapolis: Bibliotheca Islamica, 1977 / Cairo: AUC Press, 1999

[*Al-Hubb Taht Al-Matar* (Love in the Rain), 1973]. *Amor bajo la lluvia*. Spanish transl. by Mercedes del Amo. Madrid: Editorial CantArabia, 1988

[*Al-Karnak*, 1974]. Translated by Saad El-Gabalawy in *Three Contemporary Egyptian Novels*. Fredericton, New Brunswick: York Press, 1979. Spanish transl. by M. Luisa Prieto: *Café Karnak*, Barcelona: Ediciones Martinez Roca, 2001

[*Hadrat Al-Muhtaram*, 1975]. *Respected Sir*. Translated by Rasheed El-Enany. London: Quartet Books, 1986 / Cairo: AUC Press, 1987 / New York: Doubleday, 1990

[*Hikayat Haratina*, (Stories of our alley) 1975]. Translated as *Fountain and Tomb* by Soad Sobhy, Essam Fattouh, and James Kenneson. Washington, D.C.: Three Continents Press, 1988, 1991

[*Malhamat Al-Harafish* (Epopee of the Harafish) 1977]. Translated as *The Harafish* by Catherine Cobham. Cairo: AUC Press, 1994 / New York: Doubleday, 1994

[*Asr Al-Hubb* (The Age of Love); 1981]. Italian transl. by Tania Dragotti and Elisabetta Landi: *Il tempo dell'amore*. Naples: Tullio Pironti Editore, 1990

[*Afrah Al-Qubba*, 1981]. *Wedding Song*. Transl. by Olive E. Kenny, edited and revised by Mursi Saad El Din and John Rodenbeck. Cairo: AUC Press, 1984 / New York: Doubleday, 1990

[*Layali Alf Layla*, 1982]. Translated as *Arabian Nights and Days* by Denys Johnson-Davies. Cairo: AUC Press, 1995 / New York: Doubleday, 1995

[*Rihlat Ibn Fattuma*, 1983]. *The Journey of Ibn Fattouma*. Transl. by Denys Johnson-Davies. Cairo: AUC Press, 1992 , 1997 / New York: Doubleday, 1992

[*Al-'A'ish fi-l-Haqiqa* (Rooted in Truth) 1985]. Translated as *Akhenaten, Dweller in Truth* by Tagreid Abu-Hassabo. Cairo: AUC Press, 1998 / New York: Anchor Books, 2000

[*Yawm qutila Al-Za'im*, 1985]. *The Day the Leader Was Killed*. Transl. by Malak Hashem. Cairo: General Egyptian Book Organisation, 1989; AUC Press, 1997 / Cairo AUC Press, 1997 / New York: Anchor Books, 2000

[*Qushtumur*, 1989]. *El café de Qúshtumar*. Spanish transl. Isabel Hervás Jávega. Barcelona: Destino, 1998

[*Asdaa Al-Sira Al-Dhatiyya*, 1994]. Translated as *Echoes of An Autobiography* by Denys Johnson-Davies; foreword by Nadine Gordimer. Cairo: AUC Press, 1996 / New York: Doubleday, 1996

SECONDARY LITERATURE

[*Al-Ahram*] "Complete Works: Arabic works of Naguib Mahfouz, published in Cairo by Maktabat Misr". 13–19 December 2001, Issue No. 564. http://weekly.ahram.org.eg/2001/564/8sc2.htm

al-Ashmawi-Abouzeid, Fawzia. *Le Femme et l'Egypte moderne dans l'oeuvre de Naguib Mahfouz, 1939–1967*. Geneva: Labor et Fides, 1985

Allen, Roger. *The Arabic Novel: A Historical and Critical Introduction*. Syracuse, NY: Syracuse Univ. Press, 1982

Arkoun, Mohammed. *La pensée arabe*, 3. éd., Presses univ. de France: Paris, 1985

Beard, Michael and Adnan Haydar (eds.). *Naguib Mahfouz: From Regional Fame to Global Recognition*, Syracuse, NY: Syracuse Univ. Press, 1993

Berque, Jacques. *Langages arabes du présent*. Éd. rev. et augm, Paris: Bibliothèque des sciences humaines, Editions Gallimard, 1980

Berque, Jacques. *L'Égypte: impérialisme et révolution*. Paris: Bibliothèque des sciences humaines, Editions Gallimard, 1967

Books on Egypt and Chaldaea Vol. 32: Egyptian Literature, Vol. I., *Legends of the Gods*. ed. E. A. Wallis Budge, London, 1912

Brugman, J. *An Introduction to the History of Modern Arabic Literature in Egypt.* Leiden: E.J. Brill, 1984

Budge, Sir E. A. Wallis (ed.). *The Book of the Dead.* London: Routledge & Kegan Paul, 1969

Egypt. Cairo: Lehnert & Landrock, 1976

El-Enany, Rasheed. *Naguib Mahfouz: The Pursuit of Meaning.* London: Routledge, 1993

Elkhadem, Saad. *History of the Egyptian Novel.* Fredericton, New Brunswick: York Press, 1985

Ghazoul, Ferial J. *Nocturnal Poetics: The Arabian Nights in Comparative Context.* Cairo: The American Univ. in Cairo Press, 1996

Ghitany, Gamal [Al-]. *Mahfouz par Mahfouz: entretiens avec Gamal Ghitany* / traduits de l'arabe par Khaled Osman. Paris: Sindbad, 1991 [Original title: *Najib Mahfuz yatadhakkar,* Cairo: Akhbar al-Yawm, 1987]

Gordon, Haim. *Naguib Mahfouz's Egypt: Existential Themes in His Writings.* New York; London: Greenwood Press, 1990

Hartmann, Martin. *The Arabic Press of Egypt.* London: Luzac & Co., 1899

Herdeck, Donald E. (ed.). *Three Dynamite Authors: Derek Walcott (Nobel 1992), Naguib Mahfouz (Nobel 1988), Wole Soyinka (Nobel 1986).* Colorado Springs, Colorado: Three Continents Press, 1995

Ibrahim-Hilmy, Prince, son of Ismail, Khedive of Egypt, 1860–1927. *The literature of Egypt and the Soudan, from the earliest times to the year 1885* [i.e. 1887] *inclusive. A bibliography, comprising printed books, periodical writings and papers of learned societies, maps and charts; ancient papyri, manuscripts, drawings, etc.* London: Trübner, 2 vols., 1886–87

Jacquard, Roland. *Fatwa contre l'Occident.* Avec la collaboration de Dominique Nasplèzes. Paris: Albin Michel, 1998

Le Gassick, Trevor (ed.). *Critical Perspectives on Naguib Mahfouz.* Washington, D.C.: Three Continents Press, 1991

Le Va, Britta. *The Cairo of Naguib Mahfouz.* Photographs by Britta Le Va; text by Gamal al-Ghitani; foreword by Naguib Mahfouz. Cairo: American Univ. in Cairo Press, 2000

Lyons, Robert (ed.). *Egyptian Time*. Photographs by Robert Lyons; short story by Naguib Mahfouz; translated by Peter Theroux; introduction by Charlie Pye-Smith. New York: Doubleday, 1992.

Mehrez, Samia. *Egyptian Writers between History and Fiction: Essays on Naguib Mahfouz, Sonallah Ibrahim, and Gamal al-Ghitani*. Cairo: American Univ. of Cairo, 1994

Moosa, Matti. *The Early Novels of Naguib Mahfouz: Images of Modern Egypt*. Gainesville, Florida: Univ. Press of Florida, 1994

Mostyn, Trevor. *Egypt's Belle Epoque: Cairo 1869–1952*. London: Quartet Books, 1989

Moussa-Mahmoud, Fatma. *The Arabic Novel in Egypt (1914–1970)*. Cairo: The Egyptian General Book Organization, 1973

Naguib Mahfouz: 90th Birtday Celebration, December 11, 2001. Includes bibliography of Arabic works and lists translations into foreign languages. Cairo; New York: The American Univ. in Cairo Press, 2001

Sadgrove, P. C. *The Development of the Arabic Periodical Press and Its Role in the Literary Life of Egypt*. Edinburgh [thesis]: Univ. of Edinburgh, 1983

Said, Edward W. "The Cruelty of Memory". *The New York Review*, Vol. XLVII (2000), No. 19

Stagh, Marina. *The Limits of Freedom of Speech: Prose Literature and Prose Writers in Egypt under Nasser and Sadat*. Acta Universitatis Stockholmiensis: Stockholm Oriental Studies 14. Stockholm: Almqvist & Wiksell, 1993

Patrick White

Non-fiction

Flaws in the Glass: A Self-portrait. London: Cape, 1981

Patrick White Speaks. Eds. Christine Flynn and Paul Brennan. Sydney: Primavera Press, 1989 / London: Jonathan Cape, 1990

Letters. Edited by David Marr. London: Jonathan Cape, 1994

Poetry

"Notebook, 1934" [manuscript], in the National Library of Australia, Canberra. Contains thirty poems written by White, and an index. Twenty-eight of the poems were subsequently published in "The Ploughman and Other Poems" (1935). The notebook was given to Elizabeth Withycombe by White.

The Ploughman and Other Poems. Illustrations by L. Roy Davies. Sydney: Beacon Press, 1935

Henderson, Moya. *Six urban songs* [music]: for mezzo-soprano & orchestra. Poems by Patrick White. Grosvenor Place, NSW: Reproduced and distributed by Australian Music Centre, 1999, c1983

Novels

Happy Valley. London: Harrap, 1939

The Living and the Dead. London: Routledge, 1941 / New York: Viking, 1941

The Aunt's Story. London: Routledge & Kegan Paul, 1948

The Tree of Man. New York: Viking, 1955

Voss. London: Eyre & Spottiswoode, 1957 / New York: Viking, 1957

Riders in the Chariot. London: Eyre & Spottiswoode, 1961

The Solid Mandala. London: Eyre & Spottiswoode, 1966

The Vivisector. London: Jonathan Cape, 1970

The Eye of the Storm. London: Jonathan Cape, 1973

A Fringe of Leaves. London: Jonathan Cape, 1976

The Twyborn affair. London: Jonathan Cape, 1979

Memoirs of Many in One by Alex Xenophon Demirjian Gray. Edited [written] by Patrick White. London: Jonathan Cape, 1986

Short Stories

The Burnt Ones. London: Eyre & Spottiswoode, 1964

The Cockatoos. Shorter Novels and Stories. London: Jonathan Cape, 1974

Three Uneasy Pieces. London: Jonathan Cape, 1988, c1987

Plays

Four Plays. ("The Ham Funeral", "The Season at Sarsaparilla", "A Cherry Soul", "Night on Bald Mountain"). London: Eyre & Spottiswoode, 1965. Reprinted in: *Collected Plays*, Vol. 1. Sydney: Currency Press, 1993

Collected Plays, Vol. 2. Sydney: Currency Press, 1994. Contents: "Big Toys" (1978); "Netherwood" (1983); "Signal Driver" (1983); "Shepherd on the Rocks"

SECONDARY LITERATURE

Berg, Mari-Ann. *Aspects of Time, Ageing and Old Age in the Novels of Patrick White, 1939–1979.* Gothenburg: Acta Universitatis Gothoburgensis, 1983

Bliss, Carolyn. *Patrick White's Fiction: The Paradox of Fortunate Failure.* Basingstoke: Macmillan, 1986

Collier, Gordon. *The Rocks and Sticks of Words: Style, Discourse and Narrative Structure in the Fiction of Patrick White.* Amsterdam: Rodopi, 1992

Critical Essays on Patrick White. Compiled by Peter Wolfe. Boston, Mass.: G.K. Hall, 1990

During, Simon. *Patrick White.* Melbourne: Oxford Univ. Press, 1996

Edgecombe, Rodney Stenning. *Vision and Style in Patrick White: A Study of Five Novels.* Tuscaloosa; London: Univ. of Alabama Press, 1990

Giffin, Michael. *Patrick White and the Religious Imagination: Arthurs' Dream.* Lewiston, NY; Lampeter: Edwin Mellen Press, 1999. Originally published as: *Arthur's Dream.* Paddington, NSW, Australia: Spaniel Books, 1996

Hansson, Karin. *The Warped Universe: A Study of Imagery and Structure in Seven Novels by Patrick White.* Malmö: Liber/Gleerup, 1984

Laigle, Geneviève. *Le sens du mystère dans l'oeuvre romanesque de Patrick White.* Paris: Didier, 1989

Marr, David. *Patrick White: A Life.* London: Jonathan Cape, 1991

Porat, Zephyra. "'The Madwoman in the Garden': Post-Nietzschean Ethics of Interlocution in Patrick White's 'Aunt's story'". In: *The Yearbook of Research in English and American literature*, 1993, pp. 270–301

Roberts, Sheila Valerie. *The Experience of Time in the Novels of Patrick White [Die ervaring van tyd in die romans van Patrick White]*. Diss., Pretoria, 1977

Tacey, David. *Patrick White: Fiction and the Unconscious*. Melbourne: Oxford Univ. Press, 1988

Williams, Mark. *Patrick White*. New York: St. Martin's Press, 1993

Ernest Hemingway

WORKS

Three Stories & Ten Poems. Paris: Contact Editions, 1923. *Three Stories & Ten Poems: Ernest Hemingway's First Book*. A facsimile of the original Paris Edition published in 1923. Bloomfield Hills, Michigan: Bruccoli Clark Books, 1977

In our time. Paris: Three Mountains Press, 1924; *In Our Time*. New York: Boni and Liveright, 1925

The Torrents of Spring: A Romantic Novel in Honor of the Passing of a Great Race. New York: Charles Scribner's Sons, 1926

The Sun Also Rises. New York: Charles Scribner's Sons, 1926, 1928

Fiesta. London: Jonathan Cape, 1927 / Pan Books, 1972. The British version of *The Sun Also Rises*

Men without Women. New York: Charles Scribner's Sons, 1927

A Farewell to Arms. New York: Charles Scribner's Sons, 1929

Death in the Afternoon. New York: Charles Scribner's Sons, 1932 / London: Jonathan Cape, 1932

Winner Take Nothing. New York: Charles Scribner's Sons, 1933

Green Hills of Africa. New York: Charles Scribner's Sons, 1935

To Have and Have Not. New York: Charles Scribner's Sons, 1937

The Fifth Column and the First Forty-nine Stories. New York: Charles Scribner's Sons, 1938

For Whom the Bell Tolls. New York: Charles Scribner's Sons, 1940

Across the River and Into the Trees. New York: Charles Scribner's Sons, 1950

The Old Man and the Sea. New York: Charles Scribner's Sons, 1952

The Snows of Kilimanjaro and Other Stories. New York: Charles Scribner's Sons, 1961

A Moveable Feast. New York: Charles Scribner's Sons, 1964 / London: Jonathan Cape, 1964

By-Line: Ernest Hemingway. Selected Articles and Dispatches of Four Decades. Edited by William White, with commentaries by Philip Young. New York: Charles Scribner's Sons, 1967 / London: Collins, 1968

Islands in the Stream. New York: Charles Scribner's Sons, 1970

The Nick Adams Stories. Preface by Philip Young. New York: Charles Scribner's Sons, 1972

Complete Poems. Edited with an introduction and notes by Nicholas Gerogiannis. Lincoln: Univ. of Nebraska Press, 1983, c1979

Selected Letters, 1917–1961. Ed. Carlos Baker. New York: Charles Scribner's Sons, 1981 / London: Panther Books / Granada Publishing, 1985

The Dangerous Summer. New York: Charles Scribner's Sons, 1985

The Garden of Eden. New York: Charles Scribner's Sons, 1986

The Complete Short Stories of Ernest Hemingway, The Finca Vigia Edition. New York: Charles Scribner's Sons, 1987, 1998

True at First Light. Edited with an Introduction by Patrick Hemingway. New York: Charles Scribner's Sons, 1999 / London: Arrow Books/ Random House, 1999

SECONDARY LITERATURE

Baker, Carlos. *Hemingway: The Writer as Artist*. Fourth edition, Princeton, NJ: Princeton Univ. Press, 1972

Bruccoli, Matthew J. (ed.). *Ernest Hemingway's Apprenticeship: Oak Park, 1916–1917*. Washington, D.C.: NCR Microcard Editions, 1971

Bruccoli, Matthew J., and Robert W. Trogdon (eds.). *The Only Thing That Counts: The Ernest Hemingway–Maxwell Perkins Correspondence 1925–1947*. New York: Charles Scribner's Sons, 1996

Clifford, Stephen P. *Beyond the Heroic "I": Reading Lawrence, Hemingway, and "Masculinity"*. Cranbury, NJ: Bucknell Univ. Press, 1999

Josephs, Allen. *For Whom the Bell Tolls: Ernest Hemingway's Undiscovered Country.* New York: Twayne, 1994

Lacasse, Rodolphe. *Hemingway et Malraux: destins de l'homme.* Profils; 6: Éditions Cosmos: Montréal, 1972

Lynn. Kenneth S. *Hemingway.* London: Simon & Schuster, 1987

Mandel, Miriam B. *Reading Hemingway: The Facts in the Fictions.* Metuchen, NJ; London: Scarecrow Press, 1995

Meyers, Jeffrey. *Hemingway: A Biography.* New York: Harper & Row, 1985 / London: Macmillan, 1986

Nelson, Gerald B. and Glory Jones. *Hemingway: Life and Works.* New York: Facts On File Publications, 1984

Palin, Michael. *Hemingway's Travels.* London: Weidenfeld & Nicolson, 1999

Phillips, Larry W. (ed). *Ernest Hemingway on Writing.* London: Grafton Books, 1986 (1984)

Reynolds, Michael S. *Hemingway: An Annotated Chronology: An Outline of the Author's Life and Career Detailing Significant Events, Friendships, Travels, and Achievements.* Omni Chronology Series, 1. Detroit, MI: Omnigraphics, 1991

Reynolds, Michael S. *Hemingway's First War: The Making of A Farewell to Arms.* New York and Oxford: Basil Blackwell, 1987 (Princeton Univ. Press, 1976)

Reynolds, Michael S. *Hemingway: The Final Years.* New York: W.W. Norton, 1999

Reynolds, Michael S. *Hemingway: The Homecoming.* New York: W.W. Norton, 1999

Reynolds, Michael S. *Hemingway: The Paris years.* New York: W.W. Norton, 1999

Reynolds, Michael S. *The Young Hemingway.* New York: W.W. Norton, 1998

Trogdon, Robert W. (ed.). *Ernest Hemingway: A Documentary Volume.* In: *Dictionary of Literary Biography* (series) Vol. 210. Detroit, MI: Gale Research Inc., 1999

Wagner-Martin, Linda (ed.). *A Historical Guide to Ernest Hemingway.* New York and Oxford: Oxford Univ. Press, 2000

The John F. Kennedy Library in Boston, Massachusetts, has an extensive collection of books and manuscripts, and holds more than 10,000 photos of Ernest Hemingway.

Grazia Deledda

WORKS

Collections and annotated editions

Canne al vento. Ed. Lucia Genovese and Elisabetta Erre. [Annotated and illustrated edition with a glossary.] Milano: Sedes, 1993

Fabie e Leggende. Ed. Bruno Rombi. Milan: Rusconi, 1994

Leggende sarde. Ed. Dolores Turchi. Rome: Tascabelli Economici Newton, 1999

Novelle. Ed. Giovanna Cerina. 6 vols. Nuoro: Ilisso Edizioni, 1996 [coll. Bibiliotheca Sarda]. This edition contains the following collections of short stories:

Vol. I: *Nell'azzurro* (1890); *Racconti sardi* (1894); *L'ospite* (1897); *Le tentazioni* (1899)

Vol. II: *La regina delle tenebre* (1902); *I giuochi della vita* (1905); *Amori moderni* (1907); *Il nonno* (1908)

Vol. III: *Chiaroscuro* (1912); *Il fanciullo nascosto* (1915)

Vol. IV: *Il ritorno del figlio* (1919); *La bambina rubata* (1919); *Il flauto nel bosco* (1923); *Il sigillo d'amore* (1926)

Vol. V: *La casa del poeta* (1930); *Il dono di natale* (1930); *La vigna sul mare* (1932)

Vol. VI: *Sole d'estate* (1933); *Il cedro del libano* (1939)

Opere scelte. 2 vols. Ed. Eurialo De Michelis. Milan: Mondadori, 1964

Romanzi e novelle. Ed. Natalino Sapegno. Milan: Mondadori, 1971

Tradizioni popolari di Nuoro / Grazia Deledda; presentazione di Francesco Alziator e Fernando Pilia (ed. anastatica da La rivista delle tradizioni popolari italiane / diretta da Angelo De Gubernatis). Cagliari: 1972

Versi e prose giovanili. Ed. Antonio & Carmen Scano. Milan: Edizioni Vigilio, 1972

Italian pocket editions

In present-day Italy, a dozen of Grazia Deledda's novels are kept in print in fine inexpensive pocket volumes, many of them published by the Mondadori publishers, Milan, in the series Scrittori del Novecento. For example:

La via del male [1896]. Introduzione di Anna Dolfi. Milano: A. Mondadori, 1983 / Introduzione di Dolores Turchi. Ed. integrale. Roma: Tascabili economici Newton, 1994

Elias Portolu [1903]. Edited with an introduction, a collection of criticism, and a bibliography by Vittorio Spinazzola. Milan: Oscar Mondadori, 2002 (1970–)

Cenere [1903]. Ed. with an introduction and a bibliography by Vittorio Spinazzola. Milan: Arnaldo Mondadori Editore, 2001 (frequently republished and reprinted since 1961)

La Madre [1920]. Contains an introduction, a collection of criticism, and a bibliography by Vittorio Spinazzola. It also includes an analysis of the style and structure of the novel. Milan: Arnaldo Mondadori Editore, 2002 (1941–)

La chiesa della solitudine [1936]. With an introduction by Vittorio Spinazzola and an appendix by E. Ann Matter. Milan: Arnaldo Mondadori Editore, 2001 (1956–)

Cosima [1937]. With an introduction and a bibliography by Vittorio Spinazzola, and an appendix by Antonio Baldino. Milan: Arnaldo Mondadori Editore, 2001 (1947–)

Secondary Literature

Agus, Serafino. *Ipotesi di lettura di Grazia Deledda.* Dolianova (Cagliari): Grafica del Parteolla, 1999

Balducci, Carolyn F. *A Self-made Woman: Biography of Nobel-Prize-Winner Grazia Deledda.* Boston: Houghton Mifflin, 1975

Capuana, Luigi. *Gli "ismi" contemporanei. Verismo, simbolismo, idealismo, cosmopolitismo ed altri saggi di critica letteraria ed artistica.* [Catania 1898]. Ed. Giorgio Luti. Milano: Fratelli Fabbri, 1973

Cara, Antonio. *Cenere di Grazia Deledda nelle figurazioni di Eleonora Duse.* Nuoro: Istituto Superiore Regionale Etnografico, 1984

Cirese, Alberto Mario. *Intellettuali, folklore, istinto di classe: note su Verga, Deledda, Scotellaro, Gramsci.* Turin: G. Einaudi, 1976

Corda, Francesco. *Grammatica moderna del sardo logudorese.* Cagliari: Edizioni della Torre 1994 [on Grazia Deledda's native tongue]

Croce, Benedetto. *La letteratura della nuova Italia.* In: *Scritti di storia letteraria e politica* (several editions). Bari: 1912–1954

De Giovanni, Neria. *Il peso dell'eros: mito ed eros nella Sardegna di Grazia Deledda.* Alghero (Sassari): Nemapress, 2001

De Giovanni, Neria (ed.). *Religiosità, fatalismo e magia in Grazia Deledda.* Turin: Edizioni San Paolo, 1999

Dolfi, Anna. *Grazia Deledda.* Milan: Mursia, 1979

Giacobbe Harder, Maria. *Grazia Deledda: introduzione alla Sardegna.* Milan: Bompiani, 1973

Kozma, Janice M. *Grazia Deledda's Eternal Adolescents: The Pathology of Arrested Maturation.* Madison, NJ: Fairleigh Dickinson Univ. Press, 2002

Lampo, Giovanna. *Grazia Deledda verista?* Cagliari: Arte Duchamp, 2002

Lawrence, D. H. *Sea and Sardinia.* Intro. R. Aldington. London: Heineman, 1952. D. H. Lawrence's critical foreword to Deledda's novel *The Mother*, which appeared in the English editions of the 1920s, is reprinted in the new edition of M. G. Steegman's translation: *La Madre (The Woman and the Priest)* or *The Mother*, edited with an introduction and chronology by Eric Lane. London: Daedalus/Hippocrane, 1987

Olla, Gianni et al. *Scenari Sardi: Grazia Deledda tra cinema e televisione.* [Con il soggetto scenario sardo per il cinema di Grazia Deledda, 1916.] Cagliari: Aipsa, 2001

Pellegrino, Angelo (ed.). *Metafora e biografia nell'opera di Grazia Deledda.* Roma: Inst. della Enciclopedia Italiana, 1990

Piano, Maria Giovanna. *Onora la Madre: Autorità femminile nella narrativa di Grazia Deledda.* Turin: Rosenberg & Sellier, 1998

Piromalli, Antonio. *Grazia Deledda.* Firenze: La nuova Italia, 1968

Pittalis, Paola. *Storia della Letteratura en Sardegna.* Cagliari: Edizioni Della Torre, 1998

Rasy, Elisabetta. *Ritratti di signora* [on Grazia Deledda, Ada Negri and Matilde Serao]. Milano: Rizzoli, 1995

Sacchetti, Lina. *Grazia Deledda: Ricordi e testimonianze*. Bergamo: Minerva italica, 1971

Vittorini, Domenico. *High Points in the History of Italian Literature*. New York: David McKay, 1958

Amartya Sen

WORKS

Africa and India: What Do We Have to Learn from Each Other? Helsinki, Finland: WIDER, 1988

"Amartya Sen Talks with Bina Agarwal, Jane Humphries and Ingrid Robeyns". In: *Feminist Economics*, 2003(9): 2/3, pp. 319–332

Choice of Techniques: An Aspect of the Theory of Planned Economic Development, 3rd ed. New York: A.M. Kelley, 1968

Choice, Welfare and Measurement. Repr. Oxford: Blackwell, 1983

Collective Choice and Social Welfare. San Francisco: Holden Day / London: Oliver and Boyd, 1970

Commodities and Capabilities. New Delhi: Oxford Univ. Press, 1999

Development as Freedom. [New edition] Oxford: Oxford Univ. Press, 2001

On Economic Inequality: The Radcliffe lectures, delivered in the University of Warwick, 1972. Oxford: Clarendon, 1973

On Ethics and Economics. Oxford; New York: Basil Blackwell, 1987

Gender and Cooperative Conflicts. Helsinki, Finland: World Institute for Development Economics Research of the United Nations Univ., 1988

India: Economic Development and Social Opportunity / Jean Drèze, Amartya Sen. Oxford: Clarendon, 1998

Inequality Reexamined. New York: Russell Sage Foundation; Cambridge, Mass.: Harvard Univ. Press, 1992

The Political Economy of Hunger. Edited by Jean Drèze and Amartya Sen. Oxford: Clarendon Press; New York: Oxford Univ. Press, 1990–1991

Poverty and Famines: An Essay on Entitlement and Deprivation. Oxford: Clarendon Press, 1981

The Quality of Life. Edited by Martha Nussbaum and Amartya Sen. Oxford: Clarendon Press; New York: Oxford Univ. Press, 1993

Rationality and Freedom. Cambridge, Mass.; London: Belknap Press of Harvard Univ. Press, 2002

Reason before Identity. Oxford; New York: Oxford Univ. Press, 1999

Resources, Values, and Development. Cambridge, Mass.; London: Harvard Univ. Press, 1997

The Amartya Sen and Jean Drèze omnibus [elektronic resource] comprising poverty and famines, hunger and public action. New Delhi: Oxford Univ. Press, c1999

Utilitarianism and Beyond. Edited by Amartya Sen and Bernard Williams. Cambridge: Cambridge Univ. Press, 1982

SECONDARY LITERATURE

"A Special Issue on Amartya Sen's Work and Ideas: A Gender Perspective". Edited by Bina Agarwal et al., *Feminist Economics* 2003(9): 2/3

Choice, Welfare, and Development: A Festschrift in Honour of Amartya K. Sen. Edited by K. Basu, P. Pattanaik, and K. Suzumura. Oxford: Clarendon Press; New York: Oxford Univ. Press, 1995

Economics of Amartya Sen. Edited by Ajit Kumar Sinha and Raj Kumar Sen. New Delhi: Deep & Deep Publications, 2000

Nussbaum, Martha C. "Capabilities as Fundamental Entitlements: Sen and Social Justice". In: *Feminist Economics* 2003(9): 2/3, pp. 33–59

Welfare, Choice, and Development. Essays in Honour of Professor Amartya Sen. Edited by Biswanath Ray. New Delhi: Kanishka Publishers, 2001

Rabindranath Tagore

WORKS

Collected Poems and Plays of Rabindranath Tagore. 1st ed. London: Macmillan, 1936 / London: Papermac, 1990

The English Writings of Rabindranath Tagore. Edited by Sisir Kumar Das. 3 vols. New Delhi: Sahitya Akademi, 1994, 1996

[*Ghare Baire*, 1916] *The Home and the World* [novel]. Madras: Macmillan India, 1919 / Leipzig : Tauchnitz, 1921

[*Gitanjali*, 1912]. *Gitanjali. A Collection of Prose Translations made by the Author*. With an introduction by W. B. Yeats. London: Macmillan, 1914 (1913). — *Song Offerings*. New Translation by Joe Winter. London: Anvil Press, 2000 [1998]

Glimpses of Bengal: Selected from the Letters of Sir Rabindranath Tagore, 1885 to 1895. London: Macmillan, 1921

I Won't Let You Go: Selected Poems. Translated by Ketaki Kushari Dyson. Newcastle upon Tyne: Bloodaxe Books, 1991

Later Poems of Tagore. Translated by Aurobindo Bose. New Delhi: Orient Paperbacks, 1974

Letters from Russia. Translated from Bengali by Sasadhar Sinha. Calcutta: Visva-Bharati, 1960

Particles, Jottings, Sparks: The Collected Brief Poems. Translated with an introduction by William Radice. London: Angel, 2001

Personality. Lectures Delivered in America. Madras: Macmillan India, 1980, 1985. Originally published: London: Macmillan, 1917

Political Thinkers of Modern India. Vol. 25, *Rabindranath Tagore*. Ed. Verinder Grover. New Delhi: Deep & Deep Publications, 1993

Rabindranath Tagore: An Anthology. Edited by Krishna Dutta and Andrew Robinson. London: Picador, 1997

The Religion of Man. London: Unwin, 1931; New edition, 1961

A Rich Harvest: The Complete Tagore/Elmhirst Correspondence, & Other Writings. Edited by Kissoonsingh Hazareesingh. Stanley, Rose-Hill, Mauritius. Éditions de l'Océan Indien, 1992

Sacrifice and Other Plays. Macmillan pocket Tagore edition. Contents: "Sanyasi, or, The Ascetic", "Malini", "Sacrifice", "The King and the Queen [*Raja O Rani*, 1889]". Madras: Macmillan India, 1980, 1985

Selected Letters of Rabindranath Tagore. Edited by Krishna Dutta and Andrew Robinson. Cambridge Univ. Press, 1997

Selected Poems. Translated by William Radice. London: Penguin, 1985

Selected Short Stories. Translated with an Introduction by William Radice. Rev. edition. London: Penguin Books, 1994

Three Plays: Mukta-dhara, Natir Puja, Chandalika. Translated by Marjorie Sykes. Madras; Oxford: Oxford Univ. Press, 1950, 1983

Visions of India: Selections from the Works of Rabindranath Tagore... / Karan Singh... 2nd edition. New Delhi: Indian Council for Cultural Relations, 1983

Wit and Wisdom of Rabindranath Tagore: Being a Treasury of Several Thousand Valuable Thoughts Collected from the Speeches and Writings of the Bard of the East and Classified under about Eight Hundred Subjects. Compiled by N. B. Sen. New Delhi: New Book Society of India, 1961

SECONDARY LITERATURE

Berlin, Isaiah. "Rabindranath Tagore and the Consciousness of Nationality". In: *The Sense of Reality: Studies in Ideas and Their History.* New York: Farrar, Straus and Giroux, 1997

Chatterjee, Bhabatosh. *Rabindranath Tagore and Modern Sensibility.* New Delhi; Oxford: Oxford Univ. Press, 1996

Chatterji, Suniti Kumar. *World Literature and Tagore* ... Santiniketan: Visva-Bharati, 1971

Dutta, Krishna. *Rabindranath Tagore: The Myriad-Minded Man.* New York: St. Martin's Press, 1995; London: Bloomsbury, 1995, 1997

"Einstein and Tagore Plumb the Truth". *The New York Times Magazine,* August 10, 1930. Republished in Dutta and Robinson, *Selected Letters of Rabindranath Tagore*

Essays on Rabindranath Tagore: In Honour of D. M. Gupta. Edited by T. R. Sharma. Ghaziabad: Vimal Prakashan, 1987

Ivbulis, Viktors. *Tagore: East and West Cultural Unity.* Calcutta: Rabindra Bharati Univ., 1999

Kawabata, Yasunari. *The Existence and Discovery of Beauty.* Translated by V. H. Viglielmo. Tokyo: The Mainichi Newspapers, 1969

Khanolkar, Gangadhara Devarava. *The Lute and the Plough: A Life of Rabindranath Tagore.* Translated by Thomas Gay. Bombay: Book Centre, 1963

Kripalani, Krishna. *Rabindranath Tagore: A Biography.* London: Oxford Univ. Press, 1962

Mani, R. S. *Educational Ideas and Ideals of Gandhi and Tagore: A Comparative Study.* New Delhi: New Book Society of India, 1995

Nandy, Ashis. *The Illegitimacy of Nationalism: Rabindranath Tagore and the Politics of Self.* New Delhi; Oxford: Oxford Univ. Press, 1994

Nussbaum, Martha C. et al. *For Love of Country: Debating the Limits of Patriotism.* Edited by Joshua Cohen. Boston: Beacon Press, 1996

Rabindranath Tagore and the Challenges of Today. Edited by Bhudeb Chaudhuri and K. G. Subramanyan. Shimla: Indian Institute of Advanced Study; Calcutta: Distributed by Seagull Bookshop, 1988

Rabindranath Tagore: A Centenary Volume, 1861–1961. With an Introduction by Jawaharlal Nehru. New Delhi: Sahitya Akademi, 1961

Rabindranath Tagore: Perspectives in Time. Edited by Mary Lago and Ronald Warwick. Basingstoke: Macmillan, 1989

Radhakrishnan, S. *The Philosophy of Rabindranath Tagore.* London: Macmillan, 1918

Robinson, Andrew. *The Art of Rabindranath Tagore.* Foreword by Satyajit Ray. London: André Deutsch, 1989

Sen, Nabaneeta Dev. *Counterpoints: Essays in Comparative Literature.* Calcutta: Prajna, 1985

Swarup, Jagadish. *Human Rights and Fundamental Freedoms.* Bombay, 1975

Tagore's Home and the World: Modern Essays in Criticism. Edited by P. K. Datta. London: Anthem, 2003

Thompson, Edward. *Rabindranath Tagore: Poet and Dramatist.* London: 2nd edition, revised and reset, Oxford Univ. Press, 1948 / New York: Haskell, 1974

Nelson Mandela

WORKS

No Easy Walk to Freedom. Ed. Ruth First. London: Heinemann, 1965

The Struggle Is My Life. New York: revised, Pathfinder, 1986. Originally published as a tribute on his 60th birthday, London: International Defence and Aid Fund for Southern Africa, 1978.

Nelson Mandela Speaks: Forging a Democratic Nonracial South Africa. New York: Pathfinder, 1993 / Johannesburg: David Philip, 1994

Long Walk to Freedom: The Autobiography of Nelson Mandela. Boston; New York; Toronto; London: Little, Brown and Company, 1994

An Illustrated Autobiography. Boston; New York; Toronto; London: Little, Brown and Company, 1996

SECONDARY LITERATURE

Barber, James. *Mandela's world: The International Dimension of South Africa's Political Revolution 1990–99.* Oxford: James Currey, 2004

Benson, Mary. *Nelson Mandela: The Man and the Movement.* 2nd revised edition. London: Penguin, 1994

Bernstein, Hilda. *The World That Was Ours: The Story of Nelson Mandela and the Rivonia trial.* London: Unwin Paperbacks / SA Writers, Robert Vicat, 1989. Originally published: London: Heinemann, 1967

Bosch, Alfred. *Nelson Mandela: le dernier titan.* Paris: L'Harmattan, 1996

Callinicos, Luli. *The World that Made Mandela: A Heritage Trail.* Johannesburg: STE Publishers, 2000

Constitution of the Republic of South Africa 1996: As adopted by the Constitutional Assembly on May 8 1996 and as amended on 11 October 1996. Pretoria: Constitutional Assembly, 1998

Datta, A. K. *South Africa.* New Delhi: Indian Council for Africa, 1960

Derrida, Jacques, and Mustapha Tlili, eds. *For Nelson Mandela.* New York: Seaver Books, 1987

Dreaming of Freedom: The Story of Robben Island. Johannesburg and Bellville: Sached Books/Mayibuye Books, 1995

Freund, Bill. *Insiders and Outsiders: The Indian Working Class of Durban, 1910–1990.* Portsmouth, New Hampshire, London, and Pietermaritzburg: Heinemann, James Currey, and Univ. of Natal Press, 1995

Frost, Brian. *Struggling to Forgive: Nelson Mandela and South Africa's Search for Reconciliation.* London: HarperCollins, 1998

Gandhi, Mohandas Karamchand. *An Autobiography, or The Story of My Experiments with Truth.* London: Penguin, 1982

Gandhi, Mohandas Karamchand. *Satyagraha in South Africa. The Selected Works of Mahatma Gandhi.* Vol. 3. General editor: Shriman Narayan. Ahmedabad, 1969

Gandhi and South Africa, 1914–1948. Edited by E. S. Reddy, Gopalkrishna Gandhi. Ahmedabad: Navajivan Publ. House, 1993

Hagemann, Albrecht. *Nelson Mandela.* Translated by Lucy Stratten. Johannesburg: Fontein Books, 1996

Harwood, Ronald. *Mandela.* London: Boxtree, 1987 / New York and Scarborough, Ontario: New American Library, 1987

Johns, Sheridan and R. Hunt Davis, Jr., eds. *Mandela, Tambo and the ANC: The Struggle against Apartheid.* New York; Oxford: Oxford Univ. Press, 1991

Juckes, Tim J. *Opposition in South Africa: The Leadership of Z. K. Matthews, Nelson Mandela and Stephen Biko.* Westport: Praeger, 1995

Kathrada, A. M. *Letters from Robben Island: A selection of Ahmed Kathrada's prison correspondence, 1964–1989.* Edited by Robert Vassen. Belville, South Africa: Mayibuye Centre; East Lansing: Michigan State Univ. Press, 1999

Meer, Fatima. *Higher Than Hope: A Biography of Nelson Mandela.* [New edition] London: Hamish Hamilton, 1990

Meredith, Martin. *Nelson Mandela: A Biography.* London: Hamish Hamilton, 1997

Nasplèzes, Dominique. *Tokyo Sexwhale: a l'ombre de Mandela.* Paris: Jean Picollec, 1994

Nehru, Jawaharlal. *Mahatma Gandhi.* Bombay: Asia Publishing House, 1960

Nelson Mandela and the Rise of the ANC. Compiled and edited by Jurgen Schadeberg; photographs by Ian Berry ...[et al.]; text by Benson Dyantyi ...[et al.]. London: Bloomsbury, 1990

Ottaway, David. *Chained Together: Mandela, De Klerk, and the Struggle to Remake South Africa.* New York: Times Books, 1993

Reddy, Enoga and Fatima Meer (eds.). *Passive Resistance 1946: A Selection of Documents.* Durban: Madiba Publishers/Institute for Black Research, 1996

Sampson, Anthony. *Mandela: The Authorized Biography.* New York: Vintage Books 2000. Originally published by HarperCollins, London, and by Alfred A. Knopf, New York, 1999

Simons, Jack, and Ray Simons. *Class and Colour in South Africa, 1850–1950.* London: International Defence and Aid Fund for Southern Africa, 1983

Sparks, Allister. *Tomorrow Is Another Country: The Inside Story of South Africa's Road to Change.* New York: Hill & Wang, 1995

Tutu, Desmond (ed. John Allen). *The Rainbow People of God*. London: Doubleday, 1994

Waldmeir, Patti. *Anatomy of a Miracle: The End of Apartheid and the Birth of a New South Africa*. London: Viking, 1997

Online Sources

❋

Texts available at http://nobelprize.org/ the official web site of the Nobel Foundation, Sweden.

Sir V. S. Naipaul: "Two Worlds", Nobel Lecture 2001.
http://nobelprize.org/literature/laureates/2001/
naipaul-lecture-e.html

Sir V. S. Naipaul: "The Enigma of Arrival" (excerpt), 2001
(© V. S. Naipaul 1987).
http://nobelprize.org/literature/laureates/2001/naipaul-prose.html

Nadine Gordimer: "Writing and Being", Nobel Lecture 1991.
http://nobelprize.org/literature/laureates/1991/
gordimer-lecture.html

Per Wästberg: "Nadine Gordimer and the South African Experience",
2001.
http://nobelprize.org/literature/articles/wastberg/index.html

Jöran Mjöberg: "A Single, Homeless, Circling Satellite — Derek
Walcott, 1992 Nobel Literature Laureate", 2001.
http://nobelprize.org/literature/articles/mjoberg/index.html

Derek Walcott: "The Antilles — Fragments of Epic Memory", Nobel Lecture 1992.
http://nobelprize.org/literature/laureates/1992/walcott-lecture.html

Anders Hallengren: "Naguib Mahfouz — The Son of Two Civilizations", 2003.
http://nobelprize.org/literature/articles/mahfouz/index.html

Patrick White: Nobel Prize "Autobiography", 1973.
http://nobelprize.org/literature/laureates/1973/white-autobio.html

Karin Hansson: "Patrick White — Existential Explorer", 2001.
http://nobelprize.org/literature/articles/hansson/index.html

Anders Hallengren: "A Case of Identity — Ernest Hemingway", 2001.
http://nobelprize.org/literature/articles/hallengren/index.html

Anders Hallengren: "Grazia Deledda: Voice of Sardinia. The first Italian woman to receive the Nobel Prize in Literature", 2002.
http://nobelprize.org/literature/articles/deledda/index.html

Amartya Sen: Nobel Prize "Autobiography", 1998. *Les Prix Nobel: 1998.*
http://nobelprize.org/economics/laureates/1998/sen-autobio.html

Amartya Sen: "Tagore and His India", Nobel e-Museum 2001, *The New York Review of Books*, Vol. 44, No. 11, 1997.
http://nobelprize.org/literature/articles/sen/index.html

Anders Hallengren: "Nelson Mandela and the Rainbow of Culture", September 11, 2001.
http://nobelprize.org/peace/articles/mandela/index.html

The Authors

❄

Anders Hallengren (b. 1950) is an associate professor of Comparative Literature and a research fellow in the Department of Literature and History of Ideas at Stockholm University. He has served as consulting editor for literature at Nobel e-Museum, and as editor of the Swedish literary journals *Parnass* and *TFL.*

Dr. Hallengren has been a visiting fellow in the Department of History at Harvard University, and a visiting professor in the Department of Philosophy, University of Hawaii at Manoa. He has lectured in many countries, including the USA, Canada, England, Italy, the former Yugoslavia, Russia, China, India, and South Africa. In 1984, by then working as a travelling foreign-affairs journalist, Hallengren published a book on the postcolonial tug-of-war over Africa (*Cuba in Africa: A Turning Point in Great-Power Politics — Decolonization and Détente in Conflict*). He has published monographs on the American philosopher Ralph Waldo Emerson (*The Code of Concord*, 1994), the Swedish theologian Emanuel Swedenborg (*Gallery of Mirrors*, 1998), and the Italian poet Francesco Petrarca (*Petrarca i Provence*, 2003).

Karin Hansson (b. 1937) taught English and Swedish at secondary school level before starting her doctoral studies at Lund University. After completing her doctoral thesis, *The Warped Universe: A Study of Imagery and Structure in Seven Novels by Patrick White*, 1984, she worked as a teacher and researcher at Lund University, and since 1991 at Blekinge Institute of Technology. She was appointed professor of English literature in 1998. She has published literary studies and books, particularly in the field of postcolonial literature, for example *Sheer Edge: Aspects of Identity in David Malouf's Writing*, 1991, and *The Unstable Manifold: Janet Frame's Challenge to Determinism*, 1996.

She has been a member of the board of EASA (European Association for Studies on Australia). In 1997 she organized a conference on Joseph Conrad and edited the proceedings, *Journeys, Myths and the Age of Travel: Joseph Conrad's Era*, 1998.

Jöran Mjöberg (b. 1913) is a retired professor of comparative literature and the author of literary studies as well as books about different countries (Iceland, Mexico, USA).

His first work, *Dikt och diktatur*, appeared in the last year of World War II and was a survey of Swedish fiction and poetry that was inspired by a criticism and condemnation of Hitler and Nazi ideology (1933–1944). He has been lecturer at the University of Oslo (1947–1949) and taught Scandinavian languages and civilization at Harvard University from 1949 to 1953. Teaching at Växjö (Växjö Katedralskola) from 1953 to 1969, and at Lund University from 1969 to 1980, he has more recently published a presentation of Latin American literature (*Latinamerikanska författare*, 1988), a study on visionary poetry during different epochs (*De såg himmel och helvete*, 1994), a book about the role of architecture in fiction and drama (*Arkitektur i litteratur*, 1999), and studies on the lyrical contribution of two Swedish poets and Nobel Laureates: Erik Axel Karlfeldt and Pär Lagerkvist.

Per Wästberg was born in Stockholm in 1933. He received his Bachelor of Arts (comparative literature) from Harvard University in 1955 and PhD from Uppsala University in 1962 on a thesis of the African novel, 1945–60. He has been a critic and columnist at *Dagens Nyheter*, Sweden's main daily, since 1953. During 1976–82, he was Chief Editor of the same paper.

He was President of Swedish PEN, 1967–78 and President of International PEN 1979–86. He founded the Swedish Amnesty in 1963. He is a member of the European Academy of Arts and Sciences since 1980 and became a member of The Swedish Academy, chair no 12, in 1997. He is a member of the Nobel Committee for Literature since 1998.

Per Wästberg has published fifty books, novels, poetry and non-fiction. He made his debut as a 15-year old with a collection of short stories, *Boy with Soap Bubbles* (*Pojke med såpbubblor*) 1949. His breakthrough came in 1955 with the novel *Half of the Kingdom* (*Halva kungariket*).

His encounter with oppression and racism in Africa and the Third World is documented in *Forbidden Territory* (*Förbjudet område*), 1960, followed by *On the Black List* (*På svarta listan*), reportage, journal and political analysis on Rhodesia (Zimbabwe) and South Africa. Altogether one million copies of the books were printed in nine languages. It led to the author being prohibited from entering Rhodesia until its independence in 1980 and from South Africa until Nelson Mandela's release in 1990.

Contributing Nobel Prize laureates

Nadine Gordimer (South Africa, b. 1923). Received the Nobel Prize for Literature in 1991.

Sir Vidiadhar Surajprasad Naipaul (United Kingdom, b. 1932 in Trinidad). Received the Nobel Prize for Literature in 2001.

Amartya Kumar Sen (India, b. 1933). Winner of the Bank of Sweden Prize in Economic Sciences in Memory of Alfred Nobel 1998. Professor of Economics, Harvard University.

Derek Walcott (St. Lucia, b. 1930). 1992 Nobel Literature Laureate.

Patrick White (Australia, b. in London 1912, d. 1990). Received the Nobel Prize for Literature in 1973.

DISCLAIMER

The views and opinions of the authors do not state or reflect those of the Nobel Foundation or the Prize-Awarding Institutions. The Nobel Foundation does not warrant or assume any legal liability or responsibility for the accuracy, completeness or usefulness of any information.